Southern Space Studies

The Southern Space Studies series presents analyses of space trends, market evolutions, policies, strategies and regulations, as well as the related social, economic and political challenges of space-related activities in the Global South, with a particular focus on developing countries in Africa and Latin America. Obtaining inside information from emerging space-faring countries in these regions is pivotal to establish and strengthen efficient and beneficial cooperation mechanisms in the space arena, and to gain a deeper understanding of their rapidly evolving space activities. To this end, the series provides transdisciplinary information for a fruitful development of space activities in relevant countries and cooperation with established space-faring nations. It is, therefore, a reference compilation for space activities in these areas.

The volumes of the series are peer-reviewed.

Wayne Stephen Coetzee · Dries Putter
Editors

Drones in the African Battlespaces

Springer

Editors
Wayne Stephen Coetzee
University West
Trollhättan, Sweden

Dries Putter
Department of Strategic Studies
Stellenbosch University
Matieland, South Africa

ISSN 2523-3718 ISSN 2523-3726 (electronic)
Southern Space Studies
ISBN 978-3-031-87146-7 ISBN 978-3-031-87147-4 (eBook)
https://doi.org/10.1007/978-3-031-87147-4

This Springer imprint is published by the registered company Springer Nature Switzerland AG
The registered company address is: Gewerbestrasse 11, 6330 Cham, Switzerland

Preface

The origins of this book trace back to a single email—a simple inquiry that initiated an ambitious journey. In November 2022, I contacted the Faculty of Military Science at the University of Stellenbosch in South Africa to ask whether anyone there was studying the role of drones in African conflicts. After spending over a decade studying Sweden's security and defence sector, I wanted to return to my roots and re-engage with African security challenges. Drones presented a new security challenge that required further examination.

A response came quickly from Dries Putter, a scholar I had not met before who shared my interest in this topic. What began as a brief email soon evolved into a regular, cross-continental correspondence as we brainstormed ways to refine and focus our research. Determined to bring structure to our ideas, we drew from available conference funds and proposed hosting a workshop in Stellenbosch. Dries and I invited academics, independent researchers, journalists, and military experts, hoping that gathering people in one place might spark our desired direction. The discussions that unfolded were lively and insightful, yet, like so much in African conflict and security research, they were complex and occasionally scattered. We quickly realised that bringing clarity and coherence to such a broad and varied topic would be challenging. After the workshop, Dries and I sat down for dinner, eager to turn our ideas into something more substantial. It was then that the concept of writing a book took shape. A book could provide the depth and structure we needed and a chance to invite diverse African perspectives on the subject.

From the very beginning, we had a clear objective: to approach the topic of drones in African battlespaces from an African perspective, amplifying the voices of those who experience, study, and navigate these battlespaces. We knew that much of the existing research on African security issues came from European or American scholars, often at the expense of local voices and insights. We decided that this book should challenge that imbalance by placing African experiences and understandings at the heart of the narrative. The pages that follow represent the culmination of this process. They are a mosaic of perspectives—histories, case studies, and analyses mostly from African researchers, practitioners, and witnesses of the conflicts reshaped by drones. This book is a testament to the collaborative journey that began with a single email, guided by the conviction that Africans must tell the story of drones in Africa. It is, above all, a product of dialogue, determination, and the belief that understanding Africa's battlespaces requires more than technology and tactics; it demands the voices of the people whose lives are interwoven with these conflict landscapes.

In recent years, drones have reshaped the contours of warfare globally, and the African continent has not been an exception. As drones make headlines worldwide, Africa's rapidly evolving battlespaces offer unique insight into the promises and perils of this transformative technology. Yet, the story of drones in Africa remains largely underexplored, obscured by regional complexities, limited access to reliable data, and a shifting landscape of actors and alliances.

Writing about drones in African battlespaces is complex. The continent's political, social, and economic diversity makes it challenging to provide a cohesive narrative. Researching drone technology in these regions also poses unique challenges. Many sources are restricted, classified, unreliable, or rapidly changing. Information on drone operations is often secretive or affected by political sensitivities. Data on drone deployments is sometimes fragmented, either whispered by insiders or buried in government reports. Unlike some other regions, Africa's landscape for drone research is also shaped by limited infrastructure and ambiguous regulations. These factors create a multifaceted environment that requires a nuanced and context-sensitive approach.

African countries also display a vast array of conflict levels, with threats often crossing porous borders. In this context, irregular warfare tactics frequently dominate, ranging from insurgencies to terrorism. This creates an environment for drone utilisation that significantly contrasts with traditional warfare, complicating impact assessments. The diversity of actors further complicates the analysis, with state militaries, insurgents, private contractors, and foreign governments using drones with varying capabilities and strategic goals. Non-state actors, for instance, may deploy commercial drones for surveillance or propaganda. In contrast, more advanced militarised drones are controlled by state forces or foreign allies, creating a complex matrix of drone use that influences battlefield dynamics in distinct ways.

Environmental factors add yet another layer of complexity. Africa's diverse terrain—from vast deserts to dense forests—demands drones that can withstand extreme temperatures, dust, and humidity, influencing operational efficacy and data reliability. Repair, maintenance, and battery logistics vary significantly depending on the environment, adding critical context for understanding drones' practical impact on African conflicts. Furthermore, ethical considerations loom large. Drones are deployed for military use and humanitarian missions, such as tracking migration or delivering medical supplies, blurring the line between military and civilian applications. This dual-use scenario raises the risk of misinterpretation and heightens the need for a balanced approach, as humanitarian applications of drones in conflict zones may inadvertently escalate tensions.

Drones are fundamentally altering the dynamics of conflicts in Africa, equipping both state and non-state actors with new capabilities. Their deployment raises critical questions about sovereignty, security, ethics, and the evolving relationship between technology and power on the continent. With these shifts come significant implications—not only for military strategy but also for human rights, geopolitical stability, and the future of warfare itself. In exploring the role of drones in Africa, this book seeks to provide clarity amid the complexity. However, this work is not simply about the machinery of drones but the broader systems of power and control they enable. It is a story of innovation meeting resilience, where technology intersects with culture, politics, and economics in unforeseen ways. Ultimately, this book is a call to consider the future of warfare in Africa with urgency and care. In lifting the veil on drone use in African battlespaces, we hope to foster a deeper understanding of this technology's challenges and opportunities. As Africa stands on the frontlines of technological change in conflict, it is essential to examine, question, and understand the ramifications for the continent and beyond.

Trollhättan, Sweden — Wayne Stephen Coetzee
Matieland, South Africa — Dries Putter

Contents

Editors and Contributors

About the Editors

Wayne Stephen Coetzee is a senior lecturer in Political Science and former Director of the International Programme in Politics and Economics at the University West in Sweden. Since 2018, Coetzee has been a visiting scholar at the University of the Western Cape, South Africa. His research primarily focuses on security politics, emphasising the interactions between public and private security, transnational defence collaborations, foreign and security policy, and the global arms trade. His publications appear in international outlets such as Cambridge University Press, *Defence and Security Analysis*, and *Cooperation and Conflict*. Over the past decade, Coetzee has also actively participated in public lectures and debates, authored "popular science" articles, and provided media insights on various subjects, including the Ukraine–Russia War, US–China relations, international law, and the security risks of environmental degradation. Coetzee earned his Ph.D. from the School of Global Studies at the University of Gothenburg and holds MA and BA degrees from Stellenbosch University.

Dries Putter is a senior lecturer at the Faculty of Military Science (FMS) at Stellenbosch University. He has been an active-duty senior naval and intelligence officer in the South African National Defence Force since 1988. As the chair for Security and Africa Studies at the FMS, his interests include intelligence studies, specifically counterintelligence, grey zones, and hybrid warfare. Putter is the chair of the FMS Research Committee and a member of the Internationalisation Committee. He is part of the Steering Committee for the newly established Defence Artificial Intelligence Research Unit (DAIRU). He is also a researcher for SIGLA at the FMS (Stellenbosch University), an affiliated member of the National Security Hub (University of Canberra), and a researcher for the Africa Research Institute, Óbuda University (Hungary). His research interests encompass intelligence, counterintelligence, covert action, hybrid warfare, arms control, and the defence industry, with publications in the *International Journal of Intelligence and Counterintelligence*, *Global Policy*, and *Scientia Militaria*.

Contributors

Karen Allen Department of War Studies, King's College, London, UK; Centre for Information Integrity in Africa, University of Stellenbosch, Stellenbosch, South Africa

Danille Elize Arendse Department of Psychology, University of Pretoria, Pretoria, South Africa

Kyle John Bester University of South Africa (Unisa), Johannesburg, South Africa

Brendon J. Cannon Khalifa University, Abu Dhabi, UAE

Wayne Stephen Coetzee School of Business, Economics and IT, University West, Trollhättan, Sweden

Arinze Ngwube Department of Political Science, Federal University Oye-Ekiti, Oye, Ekiti, Nigeria

Francis Okpaleke Department of Political Science, University of Waikato, Hamilton, New Zealand

Samuel Oyewole Federal University Oye-Ekiti, Oye, Ekiti, Nigeria

Graeme Plint Faculty of Military Science, Stellenbosch University, Saldanha, South Africa

Dries Putter Faculty of Military Science, Stellenbosch University, Saldanha, South Africa

Tshegofatso Johanna Ramachela Ocean Regions Programme, University of Pretoria, Pretoria, South Africa

Ash Rossiter Khalifa University, Abu Dhabi, UAE

Warfare Reimagined: How Drones Are Shaping the African Battlespaces

Wayne Stephen Coetzee and Dries Putter

Abstract

This introductory chapter establishes the analytical and thematic foundations of the volume. It positions drone proliferation as a key element of current African security dynamics, emphasizing how state and non-state actors utilize unmanned systems across the continent for surveillance, military operations, peacekeeping, and coercive actions. The chapter presents the volume's central argument: Africa is not simply a passive site for drone deployment but an active geopolitical arena where global power struggles, security innovations, and regulatory gaps intersect. It outlines the book's structure and scope, highlighting three main themes—strategic and operational effects, contextual effectiveness, and ethical-legal issues—that inform the chapters. In this way, the introduction sets the stage for a more extensive examination of how drones are transforming the dynamics of power, sovereignty, and security governance in Africa and beyond.

1 Introduction

The rapid proliferation of unmanned systems, also known as drones, has redefined modern warfare, reshaped global security paradigms, and introduced new layers of geopolitical complexity. An unmanned system is defined as 'any electromechanical system which has the ability to carry out a predetermined or described task, or a portion of that task, and do it automatically with limited or no human

W. S. Coetzee (✉)
School of Business, Economics and IT, University West, Trollhättan, Sweden
e-mail: wayne-stephen.coetzee@hv.se

D. Putter
Faculty of Military Science, Stellenbosch University, Saldanha, South Africa
e-mail: putter@sun.ac.za

W. S. Coetzee and D. Putter (eds.), *Drones in the African Battlespaces*, Southern Space Studies, https://doi.org/10.1007/978-3-031-87147-4_1

intervention'.[1] These systems can operate in various environments, including air, land, water, and space, and are typically categorised according to their functionality: AUV (autonomous unmanned vehicles); UAVs (unmanned aerial vehicles); UAS (unmanned aircraft systems); RPAs (remotely piloted aircraft); UGS/UGVs (unmanned ground systems/vehicles); USS (unmanned surface systems); UMS (unmanned marine systems); UUS/UUVs (unmanned underwater systems/, vehicles); and ROVs (remotely operated vehicles). Although the term 'drone' is commonly linked to military-type airstrikes or recreational gadgets in lay circles, we use it here as a broad term encompassing various types of unmanned systems.

This book focuses on drones in African battlespaces, highlighting Africa as both a testing ground and a theatre of profound military, political, social, and economic transformations driven by drone technology. As is evident elsewhere, drones in Africa serve multiple purposes: (i) they are used in peacekeeping and humanitarian missions,[2] (ii) as weaponised or intelligence-gathering tools by state and non-state actors[3]; and (iii) for covert surveillance of critical infrastructure, and other intelligence operations.[4] The versatility of drones as utility enablers is only limited by human imagination—reinforcing the evolutionary nature of both the technology and the application thereof. The African battlespaces provide a vast array of opportunities in support of such evolution.

The increasing availability of drones, driven by both dual-use and military technology, is reshaping the dynamics and complexity of conflicts on the African continent and beyond. Drones introduce, amongst other things, a new asymmetry that can (and in some cases already do) prolong instability and enable irregular

[1] CTU, "Unmanned and autonomous systems: The future is now", *Capital Technology University*, 2018. www.captechu.edu/blog/unmanned-and-autonomous-systems-future-now (all websites cited in this publication were last accessed and verified on 1 December 2024).

[2] Andrews, S.M., "Drones in the DRC: A case study for future deployment in United Nations peacekeeping", *Intersect* 10, no. 2 (2017), 1–10; Karlsrud, J., "The UN at war: Examining the consequences of peace-enforcement mandates for the UN peacekeeping operations in the CAR, the DRC and Mali", *Third World Quarterly* 36, no. 1 (2015), 40–54. https://doi.org/10.1080/01436597.2015.976016; Leedom, M., "Drones deliver humanitarian aid in Africa", Council on Foreign Relations, 2024. www.thinkglobalhealth.org/article/drones-deliver-humanitarian-aid-africa

[3] Ajala, O., "The use of drones in West Africa and the Sahel", in *De Gruyter Handbook of Drone Warfare*, ed. J. Patton (Berlin: De Gruyter, 2024), 255–266. https://doi.org/10.1515/9783110742039-018; Haugstvedt, H., "A flying reign of terror? The who, where, when, what, and how of non-state actors and armed drones", *Journal of Human Security* 19, no. 1 (2023), 1–7. https://doi.org/10.12924/johs2023.19010001; Okoli, A.C. and Ngom, S.(Eds.), *Banditry and security crisis in Nigeria* (New York: Routledge, 2023). https://doi.org/10.4324/9781003350149.

[4] Abiodun, T.F., "Usage of drones or unmanned aerial vehicles (UAVs) for effective aerial surveillance, mapping system and intelligence gathering in combating insecurity in Nigeria", *African Journal of Social Sciences and Humanities Research* 3, no. 2 (2020), 29–44; Gwatiwa, T., "Intelligence, surveillance and reconnaissance in the expansion of AFRICOM", in *Expanding US Military Command in Africa*, eds. T. Gwatiwa and J. v. d. Merwe (New York: Routledge, 2020). https://doi.org/10.4324/9780429449901; Rotte, R., "Western drones and African security", *African Security Review* 25, no. 1 (2016), 85–94. https://doi.org/10.1080/10246029.2015.1124043.

warfare.[5] These technologies impact both national and international security in various ways, offering low-cost solutions that can achieve significant results without the need for extensive force deployments while also providing plausible deniability from an intelligence perspective.[6]

The use of drones in Africa is a double-edged sword. As drone warfare spreads across the continent, it opens new opportunities for global defence industries, expanding the market for drone technologies and offering substantial revenue potential for state-owned defence sectors. These developments also affect international relations, reshaping ties between states involved in the drone supply chain and those impacted by their kinetic and non-kinetic applications.[7]

Domestically, African governments may aim to develop their own industries, leveraging commercially available technologies to meet national security needs. If successful and operating outside the framework of international arms control, these industries could potentially become preferred suppliers for insurgent and terrorist groups, especially if drone development and usage in Africa remain largely unregulated.[8] Additionally, non-state actors with engineering expertise are likely to exploit the accessibility of this technology, using drones of varying sophistication to advance their asymmetric objectives, from espionage to terrorism.[9]

---

[5] Kaag, J. and Kreps, S., *Drone warfare* (Cambridge: Polity Press, 2014); Udoh, G. (2024), "Artificial intelligence and the future of armed conflicts: Contextualising Africa in the global matrix", in *Artificial Intelligence, Ethics and the Future of Warfare: Global Perspectives*, ed. K. Roy, Routledge. https://doi.org/10.4324/9781003421849.

[6] Cormac, R. and Aldrich, R.J., "Grey is the new black: Covert action and implausible deniability", *International Affairs* 94, no. 3 (2018), 477–494. https://doi.org/10.1093/ia/iiy067; Fowler, M., "The strategy of drone warfare", *Journal of Strategic Security* 7, no. 4 (2014), 108–119. www.jstor.org/stable/26465233; Kumar, A., "Drone proliferation and security threats: A critical analysis", *Indian Journal of Asian Affairs* 33, no. 1/2 (2020), 43–62. www.jstor.org/stable/27003434 (all websites cited in this publication were last accessed and verified on 1 December 2024).

[7] Garcia, E., "The technological leap of AI and the Global South: Deepening asymmetries and the future of international security", SSRN, 2022. https://ssrn.com/abstract=4304540; Sanders, G., Butchart, R., Price, A., Steinberg, D. and Holderness, A., *Rising demand and proliferating supply of military UAS: Exploring demand from new UAS importers and options for U.S. security cooperation and industrial base policy* (1st ed.; Lanham: Rowman and Littlefield Publishers; 2023). https://doi.org/10.5771/9781538170694; Steff, R., Burton, J. and Soare, S., *Emerging technologies and international security: Machines,the state,and war* (New York: Routledge; 2021).

[8] Huneberg, S., "On drones, new risks and insurance", *Journal of Contemporary Roman-Dutch Law* 80 (2018), 586–601; Kumar, "Drone proliferation and security threats: A critical analysis", 43–62; Lockhart, A., While, A., Marvin, S., Kovacic, M., Odendaal, N. and Alexander, C., "Making space for drones: The contested reregulation of airspace in Tanzania and Rwanda", *Transactions of the Institute of British Geographers* 46 (2021), 850–865; Merkert, R. and Bushell, J., "Managing the drone revolution: A systematic literature review into the current use of airborne drones and future strategic directions for their effective control", *Journal of Air Transport Management* 89 (2020), 101929. https://doi.org/10.1016/j.jairtraman.2020.101929; Stöcker, C., Bennett, R., Nex, F., Gerke, M. and Zevenbergen, J., "Review of the current state of UAV regulations", *Journal of Unmanned Vehicle Systems* 9, no.5 (2017), 1–26.

[9] Haugstvedt, H., "A flying threat coming to Sahel and East Africa? A brief review", *Journal of Strategic Security* 14, no. 1 (2020), 92–105; Olech, A., "Unmanned aerial vehicle—a lethal weapon

2 Volume Scope

This edited volume explores the multifaceted impact of drones on African security landscapes, emphasising the intersection of technology, strategy, and power in a rapidly changing global context. Our central argument is that the growing reliance on drone technology in Africa carries far-reaching implications for military effectiveness, geopolitics, and the balance of power between external powers and African states. While drones provide tactical advantages, their deployment also presents significant challenges related to sovereignty, ethics, and security, especially as both state and non-state actors increasingly adopt and weaponise this technology. Each chapter in this volume offers insights into how various African states, external actors, and non-state actors employ drones in the region, reflecting on their operational efficacy, legal implications, and broader impact on African geopolitics. To that end, three overarching topics guide the investigation in this volume:

- The strategic, operational, and geopolitical impacts of drone proliferation in African battlespaces.
- The effectiveness of drones in addressing the unique security challenges posed by Africa's vast, diverse, and challenging environments.
- The ethical, legal, and socio-political concerns that arise from the increased use of drones by state and non-state actors in Africa.

3 Chapter Overview

Graeme Plint and Dries Putter, in Chap. 2, explore a conceptual grounding for understanding the broader typologies of unmanned systems, advocating for a shift away from platform-centric discussions towards a more holistic understanding of unmanned systems. While primarily seen as aerial, they argue that drones are part of a wider "unmanned" and automated battlespace that includes maritime and land-based systems. This broader perspective is critical for understanding how drones interact with other military technologies, especially in regions like Africa, where operational environments pose unique challenges.

From a geopolitical standpoint, Africa is not just a battleground for drone warfare but a lucrative market and sphere of influence for external actors. The concept of 'drone colonialism' is explored in Chap. 3 by Francis Okpaleke and Sam Oyewole, who suggest that drone proliferation in Africa is not merely a tactical development but part of a broader strategy by powerful states such as the United States, China, Iran and Turkey to maintain control over African airspaces and resources. This neo-colonial posture, reinforced by the export of military technologies, raises important questions about the long-term impact of drone dependency

of tomorrow for terrorists", *Nowa Polityka Wschodnia* 32, no. 1 (2022), 44–60; Steff, Burton, and Soare, *Emerging technologies and international security: Machines, the state, and war*.

on African sovereignty and political stability. However, the authors refine their argument by emphasising that this process is not exclusively driven by external actors. African states are increasingly acquiring drones for internal security purposes, and this surge in drone acquisition has enhanced Africa's counterterrorism capabilities. The chapter concludes by underscoring the need for African leaders to craft nuanced policies that address the long-term implications of drone proliferation on the continent's political stability and security.

In Chap. 4, Brendon Cannon and Ash Rossiter provide an analysis of the Turkish-made Bayraktar TB2 drone in Sub-Saharan African conflicts. While drones like the TB2 have proven effective in targeting insurgents and securing supply lines, their utility is often blunted by the region's vast distances, challenging environments, and the lack of sophisticated air defence systems. The authors argue that, despite their tactical advantages in some contexts, these MALE (Medium Altitude Long Endurance) drones face limitations due to the absence of sufficient operational support and infrastructure as well as the enormous, complex terrains like the Sahel, where environmental and logistical challenges diminish their operational reach.

One of the central themes explored in this volume is the use of drones in counterterrorism operations, as outlined in Chap. 5 by Arinze Ngwube. This chapter focuses on France's use of drones in Mali, particularly through Operation Barkhane. It exemplifies how drones, such as the American-made MQ-9 Reaper, have become pivotal tools in monitoring vast territories, gathering intelligence, and conducting precision strikes while minimising the risk to ground troops. These operations are part of a broader counterterrorism strategy to combat groups such as Al-Qaeda in the Islamic Maghreb (AQIM) and the Islamic State in the Greater Sahara (ISGS). However, France's reliance on drone-based counterinsurgency operations also raises ethical concerns regarding civilian casualties and the sovereignty of African states. This case demonstrates both the operational success and the moral dilemmas surrounding the use of drones in warfare.

Kyle Bester and Danille Elize Arendse examine, in Chap. 6, the broader discourse surrounding drones as emerging security threats, focusing on the political, ethical, and operational challenges that come with integrating unmanned aerial vehicles into national security frameworks in South Africa. The chapter maps how South Africa's Armed Forces have integrated drones into their security strategy, using unmanned aerial vehicles for surveillance and intelligence gathering. Yet, the authors raise concerns about the privacy, safety, and potential misuse of drones by non-state actors, particularly given the potential for cyberattacks and the accessibility of drone technology. The securitization of unmanned aerial vehicles, framed through the lens of Securitization Theory, illustrates how drones are increasingly seen as both tools of national security and sources of potential threats.

Drones are also increasingly being used in maritime security, providing valuable solutions for African coastal states dealing with piracy, illegal fishing, and transnational crime. As African navies and coastguards incorporate these technologies into their operations, drones present both significant opportunities and notable risks.

Their ability to monitor vast maritime territories, gather real-time data, and protect critical infrastructure like ports and shipping lanes demonstrates their potential as force multipliers in Africa's maritime domain. In Chap. 7, Tshegofatso Johanna Ramachela delves into this subject, discussing how these advancements reveal gaps in existing infrastructure and underscore the urgent need for regulatory frameworks to effectively manage drone operations. Her analysis is particularly relevant as coastal states like Nigeria and Seychelles are increasingly investing in drone technology to enhance their naval capacities. Maritime drones provide a cost-effective, high-reward solution for surveillance and security in Africa's expansive maritime areas, which are vital for international trade yet face persistent threats from piracy and illegal fishing. Ramachela explores the growing interest in drones among African littoral states, emphasizing how these technologies are being deployed to address key security challenges and supplement traditional naval forces. While drones offer considerable promise as force multipliers in Africa's often poorly monitored maritime spaces, she also highlights the technological and financial obstacles that must be overcome for their full operational potential to be realised.

The adaptation of drone technology by violent non-state actors, such as Boko Haram, al-Shabaab, and other affiliates of ISIS and al-Qaeda, represents one of the most urgent issues in African security today. The weaponization of commercially available drones and the spread of technical expertise across terrorist networks emphasize the darker side of drone proliferation. In Chap. 8, Karen Allen illustrates the complexities this trend adds to Africa's already fragile security landscape. Although insurgent use of drones in Africa is not as widespread as in the Middle East, the growing access to more sophisticated drone technologies poses significant risks. Allen traces the flow of Unmanned Aerial Systems (UAS) from terrorist organizations in the Sahel and Horn of Africa, showing how these insurgents modify drone technology for various tactical purposes. Her chapter highlights that the widespread availability of hobbyist drones, combined with open-source knowledge, has allowed these groups to adapt drones for surveillance, propaganda, and even direct attacks, creating new security challenges for African states already dealing with asymmetric warfare.

In the concluding chapter, Wayne Stephen Coetzee and Dries Putter present seven policy-relevant recommendations for African governments derived from the insights of this volume's chapters and related research. These recommendations encompass practical steps and expected outcomes.

Collectively, the chapters in this volume offer diverse perspectives on the strategic, ethical, and geopolitical implications of drone proliferation in Africa. Through case studies addressing both state and non-state uses of drones, this collection provides significant insights into how drones are transforming the African security landscape. As drones continue to play a pivotal role in modern conflict and security, it is vital to engage in critical discussions about their future applications, regulatory frameworks, and the broader consequences for peace, stability, and development.

Wayne Stephen Coetzee is a senior lecturer in Political Science and former Director of the International Programme in Politics and Economics at University West in Sweden. Since 2018, Coetzee has been a visiting scholar at the University of the Western Cape, South Africa. His research primarily focuses on security politics, emphasising the interactions between public and private security, transnational defence collaborations, foreign and security policy, and the global arms trade. His publications appear in international outlets such as Cambridge University Press, *Defence & Security Analysis*, and *Cooperation and Conflict*. Over the past decade, Coetzee has also actively participated in public lectures and debates, authored "popular science" articles, and provided media insights on various subjects, including the Ukraine-Russia War, US-China relations, international law, and the security risks of environmental degradation. Coetzee earned his PhD from the School of Global Studies at the University of Gothenburg and holds MA and BA degrees from Stellenbosch University.

Dries Putter is a senior lecturer at the Faculty of Military Science (FMS) at Stellenbosch University. He has been an active-duty senior naval and intelligence officer in the South African National Defence Force since 1988. As the Chair for Security and Africa Studies at the FMS, his interests include intelligence studies, specifically counterintelligence, grey zones, and hybrid warfare. Putter is the Chair of the FMS Research Committee and a member of the Internationalisation Committee. He is part of the Steering Committee for the newly established Defence Artificial Intelligence Research Unit (DAIRU). He is also a researcher for SIGLA at the FMS (Stellenbosch University), an affiliated member of the National Security Hub (University of Canberra), and a researcher for the Africa Research Institute, Óbuda University (Hungary). His research interests encompass intelligence, counterintelligence, covert action, hybrid warfare, arms control, and the defence industry, with publications in the *International Journal of Intelligence and Counterintelligence, Global Policy, and Scientia Militaria.*

Towards a General Theory of Unmanned Systems: A Critical Review of Drone Typology

Graeme Plint and Dries Putter

Abstract

Currently, unmanned systems are categorised between domain issues and limited operating characterisation. This multiplicity of categorisations has led to the fragmentation of knowledge between aerial systems and unmanned systems operating in other domains of war, and secondly, a focus on the concrete characteristics of unmanned aerial platforms at the expense of less observable features such as purpose and cyber issues. Lastly, the focus on the platform itself de-emphasizes important considerations that are embedded into the system in which the platform exists. This chapter proposes that a general discussion of theoretical frameworks so as to enable a broader conceptualisation of Unmanned Systems as a phenomenon. The authors argue that unmanned systems should be conceptualised broadly to include all domains (of war) and to conceptualise the typology at the system rather than platform level. Within such a broad framework, multiple characterisation points can be accommodated to identify cross-cutting similarities between unmanned systems. At the same time, the unmanned space can be achieved beyond the platform.

G. Plint · D. Putter (✉)
Faculty of Military Science, Stellenbosch University, Saldanha, South Africa
e-mail: putter@sun.ac.za

G. Plint
e-mail: plint@sun.ac.za

W. S. Coetzee and D. Putter (eds.), *Drones in the African Battlespaces*, Southern Space Studies, https://doi.org/10.1007/978-3-031-87147-4_2

1 Introduction

The rapid proliferation of drones over the past two decades in both the civil (commercial) and military spheres has changed the balance of airpower on the battlefield. Suming and Weicheng (2019) write that advancements in aerial vehicles have progressed rapidly after the historic achievement of the Wright Brothers' inaugural flight. The complexities, high costs, and inefficiencies associated with designing human-crewed vehicles (both aerial and otherwise) have spurred a significant surge in the development of unmanned alternatives, such as unmanned aerial vehicles (UAVs) and unmanned underwater vehicles (UUVs), which have evolved distinctly and rapidly over the past few decades. These changes are keenly felt within the African civil and military aerospace.

Where the civil use of drones has been well publicised, African militaries, in contrast, have been secretive about the adoption and use of drones; the full extent of adoption is difficult to assess. However, the successful use of drones in the recent conflicts in both Africa and Europe, such as Ukraine, Libya and Ethiopia, has only increased the uptake of drones in African armies. The low threshold for the use of force by African armies in counter-insurgency operations has also provided evidence of the utility of drones as a cost-effective force multiplier, giving further incentive for the adoption of drones as both sensors and effectors in the African battlespace.[1] Within the African battlespace, nationally mandated militaries are not the only actors that have acquired drone technology. The ease of access to the commercial drone market has led some defence analysts to speculate that insurgent movements in Africa may already be astute users of drones as well.[2] However, focusing only on the airspace applications of unmanned systems (colloquially referred to as drones) reveals only part of the 'unmanning' and automation of the wider African battlespace.

In maritime spaces, unmanned systems have increasingly gained traction. In 2022, the US Navy combined the International Maritime Exercise (IMX) 2022 with the exercise Cutlass Express (CE), thereby giving various East African Countries operational exposure to Unmanned Maritime Vessels (UMV).[3] Similarly,

---

[1] Adebayo, A., "Armed Drone Makes a Huge Impact in Africa", *Times Aerospace*, 6 February 2023, www.timesaerospace.aero/features/defence/armed-drone-makes-a-huge-impact-in-africa; Jeffrey, J., "Lethal Drones Are a New Menace to Africa", *New African Magazine*, 3 May 2022, https://newafricanmagazine.com/28128/; Kurpershoek, R., Valdez, A. and Zwijneneburg, W., "Remote Horizons: Expanding Use and Proliferation of Military Drones in Africa", *PAX for Peace*, February 2021, https://paxforpeace.nl/media/download/PAX_remote_horizons_FIN_lowres.pdf.

[2] Africa Defence Forum, "The Dizzying Possibilities of Drones in Africa", *defenceWeb*, 20 October 2023, www.defenceweb.co.za/aerospace/aerospace-aerospace/the-dizzying-possibilities-of-drones-in-africa/ (all websites cited in this publication were last accessed and verified on 1 December 2024).

[3] Bahtic, F., "60 Nations Kick Off Largest Maritime Exercise in Middle East", *Naval Today*, 31 January 2022, www.navaltoday.com/2022/01/31/60-nations-kick-off-largest-maritime-exercise-in-middle-east/.

other African navies, such as the Nigerian Navy, Seychelles Coast Guard, South African Navy, and Ghanian Navy, are embracing both Unmanned Aerial Vehicles (UAV) and UMVs (both surface and subsurface) to address maritime domain awareness tasks. The Nigerian Navy is moving to supplement its already established UAV capability to include Unmanned Surface Vessels (USV) to complement manned naval platforms in the Niger Delta and Lake Chad. Such a move will involve embracing increasingly more sophisticated and a higher level of Artificial intelligence (AI) and autonomy in unmanned systems (Lionel, 2024).[4] The Seychelles Coast Guard also leverage AI embedded in drones to supplement their maritime patrol capabilities.[5] The South African Navy's interest in a maritime UAV will broaden the current South African National Defence Forces' (SANDF) unmanned footprint.[6] The Ghana Navy is contemplating forming a drone squadron to assist in various naval tasks, such as intelligence, surveillance, reconnaissance (ISR) and search and rescue (SAR).[7] Thus, a wider lens is needed to fully encapsulate the role of unmanned systems in African battlespace. To accomplish this, a discussion about the current broad typology of UAVs is elaborated. After discussing the general criticism of the generic typology, a case is made for a shift in focus away from platform-centric conceptualisation towards systems characteristics. Thus, broadening the discussion allows for a smoother synchronisation between the aerospace typologies with the joint-, maritime- and landward understanding of unmanned systems. The unpacking of the nuances is supported by a discussion about the trends within such environments in support of the argument that unmanned typologies should focus firstly on automation and guidance and then domain-specific characterisations such as land, air and water. Published media is currently cluttered with many abbreviations that speak to what is available in the 'unmanned' platform space. This chapter will gradually unpack some of the nuances that define these abbreviations and, ultimately, the conceptual understanding of their use.

[4] Lionel, E., "Ghana Navy Partners Fly Zipline Ghana in Drone Operations", *Military Africa*, 22 November 2023, www.military.africa/2023/11/ghana-navy-partners-fly-zipline-ghana-in-drone-operations/.

[5] *Africa Defence Forum*, "Drones Prove Their Worth in Maritime Security", *ADF Magazine*, 24 July 2024, www.adf-magazine.com/2023/07/drones-prove-their-worth-in-maritime-security.

[6] Lesedi, S., "South African Navy Interested in Aquila Viour ALTO UAV", *Military Africa*, 2 December 2023, www.military.africa/2023/12/south-african-navy-interested-in-aquila-viour-alto-uav/.

[7] Lionel, "Ghana Navy Partners Fly Zipline Ghana in Drone Operations".

2 Drones Taxonomies

New innovations allow reconceptualisation of the world or any part of it. With any new idea, invention, or innovation, definitions and taxonomies are influential in conceptualising, shaping and eventually controlling the emerging field of knowledge. With unmanned systems, there have been intense debates around definitions and terms, as each stakeholder claims to the field.[8] A case in point is the term 'drone'. The term has become accepted as a layperson's word for UAVs. Typically, the word evokes images of unpiloted aircraft of various proportions and designs. The perceived low human footprint has resulted in drones being touted in Africa as the solution to developmental and access problems that have beset the continent.[9] In the same vein, African militaries have also adopted drones with lethal effects within the African battlespace. Traditionally, militaries, governments, and industries have grouped drones according to key attributes and specifications to create various taxonomies. The two most popular taxonomies group UAVs by size or means of propulsion.

2.1 Platform-Centricity and Manual Remote Control

2.1.1 Mass

Drone size provides a convenient single categorisation index and has been adopted by the North Atlantic Treaty Organization (NATO) and various governments and militaries.[10] Size correlates well with the drone's weight and payload size (i.e., what can be safely transported). The configuration of the payload between effectors (e.g. explosive devices), sensors and fuel determine the drone's other characteristics, such as range, speed, altitude, and endurance. These are illustrated in Table 1.

Of course, several platforms supplied by the international community fall within the description of the NATO Uncrewed Arial System (UAS) Classification (Table 1 classification). Several platforms manufactured on the African continent can also be classified according to the classification system. For example—in Africa, the Milkor 380 is an example of how Africa-based designers and manufacturers compete in the market segment between Class II and Class III of the classification system above.[11]

[8] Galbreath, D.J. and Deni, J.R. (Eds.), *Routledge Handbook of Defence Studies* (London: Routledge, 2018), 340.
[9] Africa Defence Forum, "The Dizzying Possibilities of Drones in Africa", 20 October 2023, and also see https://adf-magazine.com/2023/09/the-dizzying-possibilities-of-drones/.
[10] United States, "ATP 3-01.81 Counter-Unmanned Aircraft System (U-UAS)", *United States Army*, 2023, http://www.apd.army.mil and https://irp.fas.org/doddir/army/atp3-01-81.pdf, 1–2 and 1–3.
[11] Milkor, "Air", 2024, https://milkor.com/air-system/.

Table 1 NATO UAS classifications

NATO uncrewed arial system (UAS) classification						
Class	Category	Normal employment	Normal operating altitude	Normal mission radius	Primary supported commander	Example platform
Class III (> 600 kg)	Strike/ combat	Strategic/ national	Up to '65,000 ft	Unlimited (BLOS)	Theatre	Reaper
	High altitude long endurance (HALE)	Strategic/ national	Up to 65,000 ft	Unlimited (BLOS)	Theatre	Global hawk
	Medium altitude long endurance (MALE)	Operational/ theatre	Up to '45,000 ft MSL	Unlimited (BLOS)	JTF	Heron
Class II (150–600 kg)	Tactical	Tactical formation	Up to 18,000 ft AGL	200 km (LOS)	Brigade	Hermes 450
Class I (< 150 kg)	Small (> 15 kg)	Tactical unit	Up to 5000 ft AGL	50 km (LOS)	Battalion, regiment	Scan eagle
	Mini (<15 kg)	Tactical subunit (manual of hand launched)	Up to 3000 ft AGL	Up to 25 km (LOS)	Company, platoon, squad	Skylark
	Micro (< 66 J)	Tactical subunit (manual of hand launched)	Up to 200 ft AGL	Up to 5 km (LOS)	Platoon, squad	Black widow

Source And Szabolcsi, 2016, 195 and NATO Standard ATP-3.3.8.1 EDB V1 E[12]

2.1.2 Lift

An alternative approach is to follow the Air Force's tendency to categorise aircraft by wing type, either fixed or rotary wings (e.g. helicopters) type. The Indian government, for example, categorises UAVs into fixed-wing, rotary and hybrid

[12] Szabolcsi, R., "UAV Operator Training—Beyond Minimum Standards", *Scientific Research and Education in the Air Force* 18, no. 1 (2016), 195, https://doi.org/10.19062/2247-3173.2016.18.1.25. This table is originally from NATO Standard, "NATO Standard, "ATP-3.3.8.1 Minimum Training requirements for unmanned aircraft systems (UAS) operators and pilots", North Atlantic Treaty Organisation Allied Publication, Edition B version 1, 8 May 2019. https://nso.nato.int/nso/zPublic/ap/PROM/ATP-3.3.8.1%20EDB%20V1%20E.pdf.

(a mixture of fixed- and rotary-wing).[13] An example of a hybrid UAV can be found in a Vertical Take-off and Landing (VTOL) UAV that combine the benefits of multirotor platforms with fixed-wing drones and transition between the two modes during flight. VTOL fixed-wing drones are a versatile choice for a wide range of commercial and military aerial applications.[14] In Africa, the Narma AF200 VTOL fixed-wing UAV of Safomar Technologies is currently an example with both commercial and military applications[15]—depending on the sensors and/or kinetic capabilities fitted. Another example showcased in 2022 is the Cullinan Ximongwe, the P290e and the P400e VTOL (fixed-wing and multi-rotor).[16]

The US Army proposes three common types: fixed-wing, rotary wing/multirotor and balloons.[17] With regards to ballons, the Embry-Riddle Aeronautical University states that "Unpowered [Lighter-than-air or LTA] aircraft are considered balloons, and powered, steerable LTA aircraft are called airships [..] LTA aircraft comprise various types of balloons, airships, and other aerostats. As aeronautical technology evolved, the name "balloon" has become reserved for hot air balloons, which are recreational aircraft. While the name "aerostat" still encompasses most types of LTA aircraft, an aerostat is usually considered an unpiloted LTA aircraft tethered to the ground with a cable. The name "airship" is reserved for a piloted LTA with a propulsion system, allowing the aircraft to fly from one place to another while carrying a payload or doing something worthwhile.[18]

In the scientific research environment on the African continent, the South African Council for Scientific and Industrial Research has been developing since 2016 "what is most likely a completely unique active aerostat which it hopes to offer for security and surveillance purposes".[19] It is unique from the perspective of hybridity—the aerostat is designed with "rigid control surfaces"[20] for stabilisation and lift whilst airborne, which enhances sensor functionality and coverage,

[13] India, "The Drone Rules 2021", *The Gazette of India: Extraordinary* (2021), 324–347, https://thc.nic.in/Central%20Governmental%20Rules/Drone%20Rules,%202021.pdf.

[14] Unmanned System Technology, "Hybrid VTOL Fixed-Wing UAV Manufacturers", *Unmanned System Technology*, 2 September 2023, www.unmannedsystemstechnology.com/expo/hybrid-vtol-fixed-wing-uav-manufacturers/.

[15] Martin, G., "Safomar Showcases Innovative Narma UAV for Civil and Military Applications", *defenceWeb*, 12 March 2024, www.defenceweb.co.za/aerospace/aerospace-aerospace/safomar-showcases-innovative-narma-uav-for-civil-and-military-applications/ also see www.altiuas.com/.

[16] Droneblogger, "History as South African Company Launches Locally-Made Survey Drone", *Drone News*, 17 May 2022, https://dronenews.africa/history-as-south-african-company-launches-locally-made-survey-drone/.

[17] United States, "ATP 3-01.81 Counter-Unmanned Aircraft System (U-UAS)", 1–4.

[18] Embry-Riddle Aeronautical University, "Introduction to Aerospace Flight Vehicles—54 Airships, Blimps, & Aerostats", n.d., https://eaglepubs.eram.edu/introductiontoaerospaceflightvehicles/chapter/lth/.

[19] Martin, G., "CSIR Continues to Refine Active Aerostat", *defenceWeb*, 30 April 2019, www.defenceweb.co.za/aerospace/aerospace-aerospace/csir-continues-to-refine-active-aerostat/.

[20] Ibid.

improving the on-task time during unfavourable weather conditions., "The aerostat features an active control system, with an autopilot controlling pitch, roll and yaw. The CSIR said the active aerostat can be rapidly deployed to its required heights for maximum surveillance. Line tension is maintained within pre-set limits by the active control system".[21] Application possibilities range from "military and security, farming, wildlife monitoring, crowd control, border surveillance, mining surveillance, shark monitoring, advertising and communications relay".[22] Aerostats (at least in South Africa) allows for "persistent aerial surveillance" and an added benefit is that these capabilities do not resort under the governance requirements set for UAVs.[23]

In Africa, airspace aerostats have entered the mobile communications market segment. "World Mobile Group, the first global mobile network powered by blockchain technology and the sharing economy, successfully launches its first aerostat in Mozambique, expanding mobile connectivity to the southeast African country".[24] The longevity of such a deployment could influence decisions to deploy similar capabilities for defence and security purposes.

2.1.3 Endurance

Further characteristics used to categorise drones include power source, type of control, speed, flight duration, range, altitude, and payload.[25] However, when these taxonomies do not accommodate a novel or hybrid system, regulators are tempted to create *ad hoc* categories. For example, the European Commission (EC) created the 'Innovative Air Mobility (IAM)' category to accommodate "vehicles ranging from small drones used for cargo delivery operations to the [exotic] eVTOL aircraft"[26] and experimental vehicles such as the hybrid airship (P-791) from the Lockheed Martin stable.[27]The large scope of possible

---

[21] Ibid.

[22] Ibid.

[23] Ibid.

[24] World Mobile, "World Mobile Launches Africa's First Commercial Telecoms Aerostat, Bringing Connectivity to Rural Mozambique", *World Mobile*, 9 October 2023, https://worldmobile.io/blog/post/world-mobile-launches-africa-s-first-commercial-telecoms-aerostat-bringing-connectivity-to-rural-mozambique-1.

[25] Nader, N. and Reichert, G., "Drones in European Airspace: Initial Regulatory Steps by the EU", Centre for European Policy (cepInput) 3 (2016), www.cep.eu/fileadmin/user_upload/cep.eu/Studien/cepInput_Drohnen/cepInput_Drones.pdf, 4.

[26] Scott, B.I. and Andritsos, K.I., "A Drone Strategy 2.0 for a Smart and Sustainable Unmanned Aircraft Eco-System in Europe", Air and Space Law 48, no. 3 (2023), 278 and see European Commission, "Communication from the Commission to the European Parliament, the Council, the European Economic and Social Committee and the Committee of the Regions—'A Drone Strategy 2.0 for a Smart and Sustainable Unmanned Aircraft Eco-System in Europe'", COM(2022) 652 final and SWD(2022) 366 final, 29 November 2022, https://eur-lex.europa.eu/legal-content/EN/TXT/HTML/?uri=CELEX%3A52022DC0652.

[27] Embry-Riddle Aeronautical University, "Introduction to Aerospace Flight Vehicles—54 Airships, Blimps, & Aerostats", n.d.

attributes around which industry, governments and military develop taxonomies led Chamola, Kotesh, Agarwal, Naren, Gupta and Guizani (2021)—after considering various taxonomies—to conclude that there is still no standard unifying classification system for UAVs.[28] Broadening the understanding of drones through the development of an inclusive taxonomy—i.e. drones used in every special domain and what defines them technologically and functionally could prove very useful in understanding their rapid proliferation worldwide but in particular also into the African battle space, where the availability of legacy systems from advanced states, allow African states easy access to developing technologies.

The platform-centric classification will remain with the commercial and defence sectors for some time to come. This is due to the vast number of unsophisticated UAVs still in circulation. The cycle will be prolonged by the phasing out of less complex platforms by developed states in favour of system-centric approaches to unmanned capabilities. With the phasing out of these systems, they will be made available for procurement by developing countries such as those in Africa or just donated to countries in need (based on the strategic interests of the donor country). However, where competing international interests are at stake, the African continent will leapfrog to the system-centric approach much quicker—again, facilitated and fast-tracked by strategic international competing interests. This can typically be seen in countries with a geo-strategic position within the energy sector of Africa.

2.2 System-Centricity Progressing Towards Complete Autonomy

As platform-centric taxonomies are not unifying and do not provide sufficient description of how the platform has been unmanned or remotely controlled, a more system-centric approach could be considered. In this regard, the European Union Aviation Safety Agency (EASA) has already opted to widen the taxonomy by shifting the focus from the vehicle (which is at the platform level) to the wider unmanned system (which would include the platform and the command-and-control capability). Thus, the EASA defines Uncrewed[29] Aerial Systems (UAS) as the platform and the remote controlling equipment.[30] Another comprehensive

[28] Chamola, V., Kotesh, P., Agarwal, A., Naren, Gupta, N., & Guizani, M., "A Comprehensive Review of Unmanned Aerial Vehicle Attacks and Neutralization Techniques", *Ad Hoc Networks* 111 (2021), 2 and 3, https://doi.org/10.1016/j.adhoc.2020.102324.

[29] Uncrewed was introduced because 'unmanned' is not gender neutral. Soon they could be called They Aerial Vehicles (TAV). (Law School, "Uncrewed Military Platforms", University of Queensland, Australia, 2 October 2020, https://law.uq.edu.au/research/future-war/uncrewed-military-platforms).

[30] European Aviation Safety Agency (EASA), Easy Access Rules for Unmanned Aircraft Systems (Regulations (EU) 2019/947 and 2019/945), 2022, 20, www.easa.europa.eu/en/document-library/easy-access-rules/easy-access-rules-unmanned-aircraft-systems-regulations-eu.

description of what an uncrewed vehicle consists of can be found on the University of Queensland's (Australia) Law School webpage:

> [they can be considered as] vehicles—whether they travel on land, at sea or in the air—who do not have any people on board managing the operation of the vehicle, including the navigation, communication, flight or passage and in the take-off, landing, departure or docking, as the case might be. Some uncrewed devices can be controlled remotely by a human operator, while others might be controlled by an autonomous system. While these vehicles may have people on board, they are not involved in the operation of the vehicle itself.[31]

Re-focussing the discussion on the aerial domain—the UK Armed Forces field manual similarly juxtaposed unmanned aircraft and remotely piloted aircraft with broader definitions of unmanned aircraft systems and remotely piloted aircraft systems, thereby emphasizing the role of the support infrastructure and operating personnel in the complete unmanned system[32] (UK, 2017, p. 12)—or a system-of-systems. Such a system-of-systems would typically include an aerial vehicle to carry a payload, various support systems such as ground control stations, launch sites and recovery teams, physical and communication infrastructure and the operator.[33]

Cannon (2023) highlights the paradox—unmanning the platform does not reduce the importance of the human within the system of systems but rather accentuates the importance of the operator within such a capability. He points out that "[b]uying and operating a drone system are two different things. Armed drones are only as good as their operators, and incompetent or incorrect operation can make the best drones irrelevant".[34] Cannon argues that drones are only effective when the correct balance between numbers, payload, support infrastructure and operator is achieved.[35] Therefore, current taxonomies that focus on the platform's characteristics as an indicator of effectiveness vis-à-vis a specific application and environment require reconsideration and evolution towards system-centric approaches.

System-centric approaches involve the broad consideration of, at the very least, the data management system of the entire unmanned capability (i.e. platform and command-and-control), the navigation system and the sensor suite that facilitates autonomous or smart activities. With combat drones, consideration will also be

[31] Law School, "Uncrewed Military Platforms", University of Queensland, Australia, 2 October 2020, https://law.uq.edu.au/research/future-war/uncrewed-military-platforms.

[32] United Kingdom, JDP 0-30.2 Unmanned Aircraft Systems, Ministry of Defence, August 2017, https://assets.publishing.service.gov.uk/media/5a823670ed915d74e6236640/doctrine_uk_uas_jdp_0_30_2.pdf.

[33] Allworth, E.M., "The South African Air Force (SAAF), Unmanned Aircraft Systems and National Security: An Exploratory Study", Masters Thesis, University of Stellenbosch, 2021, 16, https://scholar.sun.ac.za.

[34] Cannon, B.J., "Out of Africa: The Impact of Drones in Sub-Saharan Conflicts", *Defense & Security Analysis* (2023), 2, https://doi.org/10.1080/14751798.2023.2178577.

[35] Ibid. 3.

given to weapon systems, which will invariably impact the real estate available for sensors and energy payloads to extend range or on-task time.

2.2.1 Data Management and Navigation

A more system-centric approach expands opportunities to discuss questions regarding data management (C^3I) within the system and increasing platform autonomy. Data communication between the platforms and the rest of UAS is critical, as the UK Armed Force JDP 0-32.2 elaborates:

> Data communication links for remotely piloted aircraft [man-in-the-loop] tend to be more complex and critical than for manned systems. Although remotely piloted aircraft are programmed to carry out a range of flight manoeuvres if control data is lost, including in extremis returning to base, they rely on a continuous stream of communications to successfully complete each mission.[36]

This example above separates platform-centric approaches to operating UAVs from system-centric approaches enabled by autonomous activity.

Chamola et al., identify four means of data communication and navigation for UAVs: (1) direct radio control, (2) satellite navigation, (3) relay control, and (4) sensor control. Direct radio control uses radio transmitters/receivers or similar electronic devices to control the platform. The nature of the transmitter restricts the UAV to a direct line of sight, thus limiting the vehicle's range. Satellite communication provides the UAV with both a data link and Geographic Positioning System (GPS) information for navigation, enabling the UAV to have greater range. Alternatively, the range of a UAV can be extended by passing data to the UAV from a series of static or moving relay stations. Finally, the UAV can be navigated in a more automated manner through internal communication (or input) from a range of onboard sensors (Chamola et al., 2021, p. 12).[37] Automated flight manoeuvre becomes increasingly more important with extended ranges and possible associated data-delivery uncertainty from radio transmitters or loss of data links due to cloud or canopy cover.

2.2.2 Sensory-Driven Autonomy

The possibility of autonomous drone manoeuvre is positively correlated to increases in onboard sensory data input, resulting in decreased reliance on human input for both navigation and decision-making. Thus, due to the complexities of unmanned aerial system-of-system configurations, UAVs can currently be ordered on a continuum between manual control and full autonomy.[38]

---

[36] United Kingdom, JDP 0-30.2 Unmanned Aircraft Systems, Ministry of Defence, August 2017, 28.

[37] Chamola, V., et al., "A Comprehensive Review of Unmanned Aerial Vehicle Attacks and Neutralization Techniques", (2021), 12.

[38] India, "The Drone Rules 2021".

2.3 Drones Within the Military Context

In the military context, UAVs can thus be defined as "an airborne vehicle which does not carry a human operator, which may be piloted remotely, follow a pre-programmed flight path, fly autonomously, or combine all three. It is designed to be recoverable and carries a lethal or non-lethal payload. Non-recoverable vehicles and projectiles such as ballistic vehicles, cruise missiles, and artillery projectiles are not considered UAVs".[39] The final part of this definition provides important distinctions for use in regulatory frameworks and international agreements that govern arms control and dual-use technology usage, for example.

2.3.1 Level of Autonomy

The UK military further grades the level of autonomy by differentiating between automated systems and fully autonomous systems, with the main difference being that automated systems respond to sensor input by following a predetermined set of rules, whereas an autonomous system behaviour is determined by the system's understanding of high-level intent, with the system selecting appropriate actions without depending on human oversight.[40] The US Navy document Unmanned Campaign Framework 2021 (Fig. 1) conceptualises the maritime-related unmanned space as follows.

With the introduction of the concept of understanding systems more towards smart systems, the above discussion illustrates that in the aerial domain, the current platform-centric classification system can benefit from a wider conceptualisation to consider system-level considerations such as data linkages (communication), navigation systems, and artificial intelligence (AI).

2.3.2 Hybridity

The inclusion of innovative platforms in the EASA categorisation suggests that the UAV typology should also be widened to include cross-domain (i.e., UAVs that can operate in any number of domains, such as on water as well as in the air or on land as well as in water) unmanned systems, and possibly expanded to conceptualise a wider joint framework that accommodates unmanned systems in other domains.

The industry is already developing unmanned vehicles that operate across domains. Such hybrid vehicles would require a combination of navigation systems to transverse between domains[41] as well as possibly different propulsion

[39] Galbreath and Deni, *Routledge Handbook of Defence Studies*, 2018, 340.

[40] United Kingdom, JDP 0-30.2 Unmanned Aircraft Systems, Ministry of Defence, August 2017, 13.

[41] Alzu'bi, H., Mansour, I. and Rawashdeh, O., "Loon Copter: Implementation of a Hybrid Unmanned Aquatic–Aerial Quadcopter with Active Buoyancy Control", *Journal of Field Robotics* 35, no. 5 (2018), 764–778; Divis, D.A., "Pegasus Multi-Domain Can Fly and Operate On the Ground", *Inside Unmanned Systems*, 20 August 2019, https://insideunmannedsystems.com/pegasus-multi-domain-can-fly-and-operate-on-the-ground/; Weisler, W., Stewart, W. anderson, M. B.,

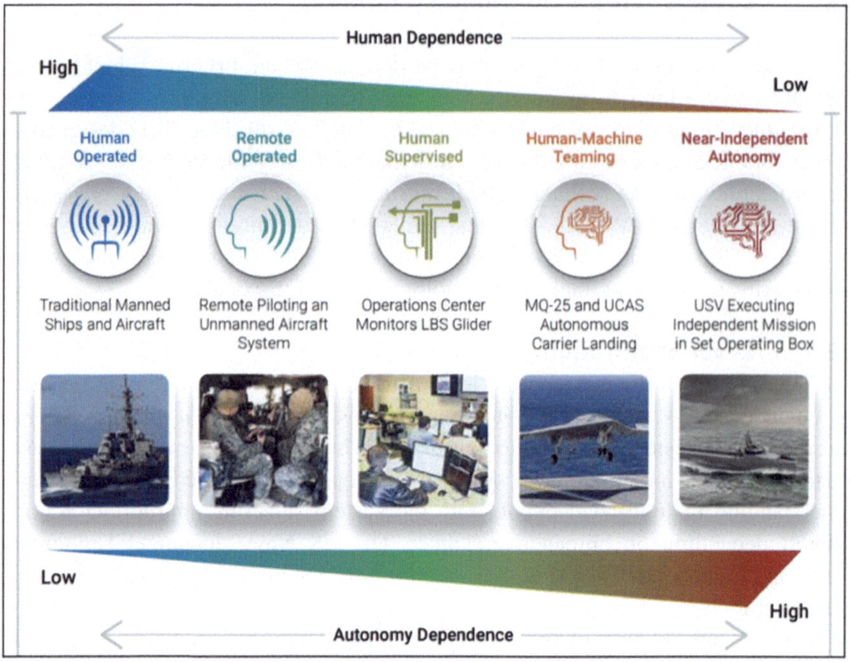

Fig. 1 US Navy document unmanned campaign framework 2021. *Source* United States, "Unmanned Campaign Framework", *US Department of the Navy*, 16 March 2021[42]

and stabilisation sub-systems. The emergence of cross-domain hybrid unmanned vehicles, such as aqua-UAVs,[43] will require the consolidation of various domain-specific taxonomies to align common characteristics. Suming and Weicheng write that aquatic UAVs are set to broaden the scope of autonomous underwater vehicles (AUVs) and UAV applications. However, designing these platforms is challenging due to "difficulties in shape design, wing design, cross-domain design, take-off and landing design, etc. and discusses some insightful solutions from the perspectives of AUVs and UAVs".[44]

Peters, K. J., Gopalarathnam, A. and Bryant, M., "Testing and Characterization of a Fixed Wing Cross-Domain Unmanned Vehicle Operating in Aerial and Underwater Environments", *IEEE Journal of Oceanic Engineering* 43, no. 4 (2018), 969–982.

[42] United States, "Unmanned Campaign Framework", *US Department of the Navy*, 16 March 2021, www.navy.mil/Portals/1/Strategic/20210315%20Unmanned%20Campaign_Final_LowRes.pdf.

[43] Yang, X., Wang, T., Liang, J., Yao, G. and Liu, M. (2015), "Survey on the Novel Hybrid Aquatic-Aerial Amphibious Aircraft: Aquatic Unmanned Aerial Vehicle (AquaUAV)", *Progress in Aerospace Sciences* 74 (2015), 131–151.

[44] Suming, Q. and Weicheng, C., "An Overview on Aquatic Unmanned Aerial Vehicles", *Annals of Reviews and Research* 5, no. 3 (2019), 10, http://dx.doi.org/10.19080/ARR.2019.05.555663.

Furthermore, within the land and maritime domain, unmanned systems have also developed at a similar pace. Integrating taxonomies across domains would be useful to deepen understanding of the unmanned system phenomenon. The added complexities of the maritime domain, such as the multi-domain nature of the seabed, subsurface, surface and aerial environments, may provide additional nuances to developing a unifying unmanned system taxonomy. Concerning the Unmanned Underwater Vehicles (UUVs), Suming and Weicheng write that "there are two main types of UUVs, i.e., AUVs and remotely operated vehicles (ROVs). UAVs and AUVs comprise a large part of unmanned vehicles. Aquatic UAVs have been developed to expand the working domains of unmanned vehicles. Easy transitions between air and water make them flexible and latent."[45]

Already in 1998, the US Department of Defence proposed a Joint Architecture for Unmanned Systems (JAUS) that transcended operating domains (i.e. land, air and water). This architecture was an engineering standard that focused on data interaction and was not a conceptual framework. Due to technology immaturity at that time, the standard did not gain traction in the industry, which led to JAUS being abandoned in 2012.[46] However, the current unmanned technology has matured sufficiently for the Australian Defence Force at the joint warfare level to adopt the conceptual classification of Robotic and Autonomous Systems (RAS). This aims at broadening the generic (non-domain specific) unmanned system concept to include automation and AI. RAS encapsulates "systems that perform a function on their own by being either physically remote from a human operator, performing cognitive-like functions on behalf of a human operator or, increasingly, both".[47]

Armies like the US and Australian Army have built on RAS to claim conceptual ownership of unmanned systems within the land and multi-domain battle space. The RAS of these armies "covers a wide range of interconnected technologies, including uncrewed systems, AI, self-learning machines, and systems more able to make sense of their environment. The increased use of RAS capabilities will continue to evolve the way Army trains and fights—enabling increased tempo, decision-making and reducing risk".[48]

Navies dealing with the challenges of a multi-domain maritime environment have also conceptualised a broader framework for unmanned systems. Bajon and Grand-Clement, faced with no universal classification system for Unmanned Maritime Systems (UMS), conceptualise UMS as systems that operate on or below the

---

[45] Ibid., 10.

[46] Breingan, C.R. and Randall, C., "Development of a System Architecture for Unmanned Systems Across Multiple Domains", Masters Thesis, Embry-Riddle Aeronautical University, 2014, 3, https://commons.eram.edu/edt/264.

[47] Vine, R. and Kohn, E., "Concept for Robotic and Autonomous Systems", Australian Defence Force, 2020, 8, https://tasdcrc.com.au/wp-content/uploads/2020/12/ADF-Concept-Robotics.pdf.

[48] Australia, "Robotic & Autonomous Systems Strategy v2.0", *Australian Army Research Centre*, August 2022, iv, https://researchcentre.army.gov.au/rico/robotic-and-autonomous-systems/robotic-autonomous-systems-ras-strategy.

water. They further distinguish UMS according to characteristics such as "size, weight (or displacement), speed, range and, for underwater systems, depth."[49] With a legacy of 'bottom-up' innovation, the US Navy faced the challenge of aligning multiple designs from multiple domains with little standardisation or commonality.[50] To bring these divergent systems into one conceptual framework, the US Navy and Marine Corps start by describing away unmanned as the "physical removal or remote location of the human from the platform" and that "Unmanned refers to the spectrum of autonomy from remote to autonomous and can be platform, system, or subsystem specific".[51] By shifting the emphasis from the platform towards the capability, the US Navy aims to implement a modular open system environment where domain-agnostic platforms can be inserted into a common operational environment with a standardised network, infrastructure, data management and analytical tools.[52] However, in developing a common open operating system with standardised communication protocols, developers must remain sensitive to each domain's unique environmental and legal considerations.[53]

Thus, by departing from the aerial platform-centric approach to focus on system decision-making, automation, and AI, the level of system autonomy becomes the key classification characteristic.[54] By focusing on autonomy-enabling attributes, unmanned systems become the catalyst for reconceptualising warfare concepts, such as a decision-centric approach to warfare. According to Clark, Patt and Schramm, "[n]ew technologies could help enable warfare concepts centred on gaining a decision-making advantage over an opponent. AI-enabled decision aids, autonomous unmanned systems, improved passive sensors, smaller weapons, and electronic and cyber warfare capabilities could impose complexity and confusion on an opponent and execute focused attacks on essential targets".[55]

---

[49] Bajon, T. and Grand-Clément, S., "Uncrewed Maritime Systems: A Primer", United Nations Institute for Disarmament Research, n.d., 1, https://unidir.org/files/2022-11/UNIDIR_Uncrewed_Maritime_Systems_Primer.pdf.

[50] Summey, D. C., Rodriguez, R. R., Demartino, D. P., Portmann, H. H., & Moritz, E., "Shaping the Future of Naval Warfare with Unmanned Systems", Dahlgren Division, Naval Surface Warfare Center, USA, July 2001, xiii/vix, https://apps.dtic.mil/sti/pdfs/ADA397057.pdf.

[51] United States, "Unmanned Campaign Framework", *US Department of the Navy*, 16 March 2021, 7.

[52] Ibid., 18–24.

[53] Rittoo, K., "Unmanned Forces: Building A Multi-Domain Autonomous Fleet", *Marine Technology News*, 26 October 2017, www.marinetechnologynews.com/news/unmanned-forces-building-multi-553750.

[54] Australia, "Robotic & Autonomous Systems Strategy v2.0", *Australian Army Research Centre*, August 2022, 6; see also Vine, R. and Kohn, E., "Concept for Robotic and Autonomous Systems", Australian Defence Force, 2020, 16.

[55] Clark, B., Patt, D. and Schramm, H., "Mosaic Warfare Exploiting Artificial Intelligence and Autonomous Systems to Implement Decision-Centric Operations", *Center for Strategic and Budgetary Assessments*, 11 February 2020, https://csbaonline.org/uploads/documents/Mosaic_Warfare.pdf.

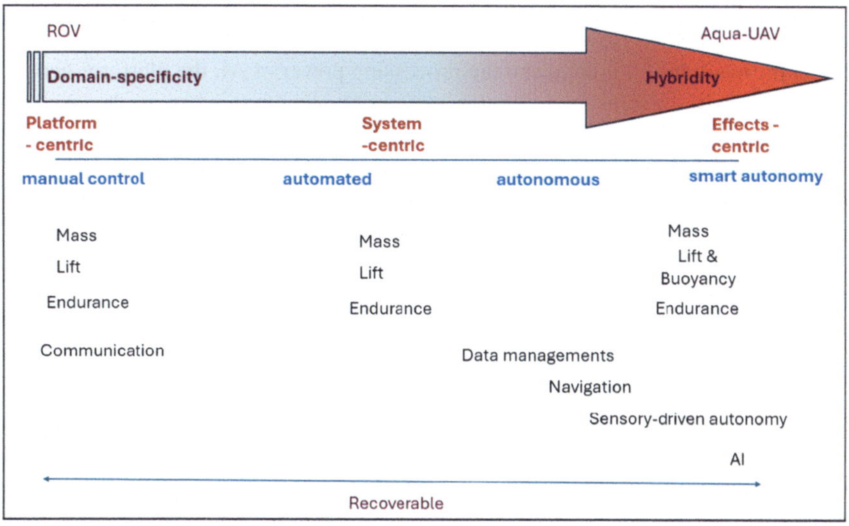

Fig. 2 Taxonomy compilation. *Source* Authors generated based on the discussion of the various taxonomies

The focus shift also allows wider discussion about integrating multiple piloted, semi-autonomous and autonomous systems into swarms. Such swarms are called Heterogeneous Autonomous Multi-Domain Swarm (HAMS). Autonomy allows multiple unmanned systems to improve "coordination, intelligence, flexibility, speed and resilience on the battlefield".[56] According to Ekelhof and Paoli, swarms are currently considered an emerging technology, requiring a robust framework to guide their operationalisation within the military domain, especially regarding aspects of human-machine interactions. Swarming fundamentally changes the current human-machine relationship common in UAV operations. As swarm sizes increase, the ability to insert a human "into the loop" of decision-making decreases, and humans are increasingly located "on the loop" rather. The focus thus moves away from controlling platforms to commanding effects or outcomes.[57]

3 Conceptual Taxonomy

The different taxonomies discussed thus far are presented in Fig. 2 to show similarities and dissimilarities.

[56] Ekelhof, M. and Paoli, G.P., "Swarm Robotics Technical and Operational Overview of the Next Generation of Autonomous Systems", *UNIDIR*, 2020, https://unidir.org/wp-content/uploads/2023/05/UNIDIR-Swarm-Robotics-2020.pdf.
[57] Ibid., 18.

Even in hybrid UAS, there will be classifications based on the platform-centric taxonomies of UAVs. However, with increasing dependence on internal communication, the availability of data, and the processing power of AI, the platform-centric taxonomies will inform operational sustainment planning, whereas the system-centric taxonomies will inform operational effects planning. As the taxonomies progress towards intelligent or smart systems, reference to these drones will probably evolve to descriptions that read as follows: Integrated land, sea, air or hybrid systems composed of AI processing and communication stations and mobile platforms that behave autonomously and self-preserving based on sensory and AI communications and processing loops; to purposefully generate effects at the tactical, operational and/or strategic levels of military action without or limited post-production human interaction.

Only human interaction with the system will be required, possibly if damaged, as well as maintenance repair and reprogramming (system updates or operational profile changes). Thus, the future of UAV/UAS taxonomies lies less in the physical characteristics of the tangible components of the system (the hardware) and more in the intangible aspects of these systems—i.e. the processing capabilities of vast amounts of real-time information provided by sensors and AI predictive capabilities. The only physical attribute that might be important is whether the UAS has a kinetic or non-kinetic mission. In this context, platform-centric taxonomies might be valuable in influencing procurement decisions.

4 Conclusion

The layperson's understanding of drones as UAVs has directed interest only towards aerial platforms. Although useful in categorising these platforms, the current taxonomy of UAVs has diverted attention away from important system attributes, such as communication, command-and-control and navigation systems. Furthermore, the air domain has attracted substantial public interest, distracting attention from equally important and mature unmanned and autonomous system development in other domains, such as land and water. However, developments here have been influenced by each specific domain's environmental and legal restrictions. In both the land and maritime domains, more focus has been placed on system autonomy and navigation than on the physical attributes of the platform—which for aerial platforms remain critical to remain airborne as required. Taking lessons from the joint environment and the maritime domain, where cross-domain integration is a significant concern, it is recommended that the current taxonomies for unmanned systems should be platform agnostic, focusing rather on higher order system level attributes such as autonomy, communication standardisation and navigation and command concerns, before grouping or describing unmanned systems by platform attributes. This approach widens the discussion to accommodate emerging technological ideas, such as swarming and provides the opportunity to align emerging warfare concepts with emerging technology.

Graeme Plint is a researcher at the Centre of Military Studies (CEMIS) at Stellenbosch University. He has been on active duty as a senior naval and intelligence officer in the South African National Defence Force since 1991. Plint has a broad interest in military studies, which include veteran affairs, youth and the military, cyber warfare, intelligence studies and naval technology. He is a member of the Faculty of Military Science Internationalisation Committee. Plint is a Ph.D. Candidate researching the Cadet Corps of the Union of South Africa.

Dries Putter is a senior lecturer at the Faculty of Military Science (FMS) at Stellenbosch University. He has been an active-duty senior naval and intelligence officer in the South African National Defence Force since 1988. As the Chair for Security and Africa Studies at the FMS, his interests include intelligence studies, specifically counterintelligence, grey zones, and hybrid warfare. Putter is the Chair of the FMS Research Committee, a member of the Internationalisation and the Teaching, Learning and Assessment Committees. He is part of the Steering Committee for the newly established Defence Artificial Intelligence Research Unit (DAIRU). He is also a researcher for SIGLA at the FMS (Stellenbosch University), an affiliated member of the National Security Hub (University of Canberra), and a researcher for the Africa Research Institute, Óbuda University (Hungary) and Advisory Board Member of the Center of Geopolitical Security and Straetgy (North West University, South Africa). His research interests encompass intelligence, counterintelligence, covert action, hybrid warfare, arms control, and the defence industry, with publications in the *International Journal of Intelligence and Counterintelligence, Journal of Policing, Intelligence and Counter Terrorism, Journal of Applied Security Research, Journal of Information Warfare, Journal of Baltic Security, Journal of Central and East European African Studies, Global Policy, African Security Review and Scientia Militaria.*

Ruling the Skies: The Geopolitics of Drone Proliferation in Africa

Francis Okpaleke and Samuel Oyewole

Abstract

The prevalence of terrorism, armed banditry, and insurgency has led to a precarious security landscape in Africa, prompting the widespread adoption of drones as crucial tools in counterterrorism and counter-banditry efforts. However, the adoption, alongside the establishment of drone bases and the proliferation of drone technology among state actors, is engendering a fresh wave of geopolitical competition among external actors. The interplay of these actors in African battlespace is reshaping supply chain dynamics, particularly in conflict-ridden regions, as they vie to establish the continent as a lucrative market and sphere of influence for drone sales. This chapter provides a focused analysis of the geopolitics surrounding drone sales and exports, with a particular emphasis on the involvement of key actors including Iran, Turkey, China, and the United States, and the spill-over effects for non-state actors. It is the argument of the chapter that the current pattern of drone proliferation for counterterrorism in Africa engenders geopolitical competition, deepens security fragilities among African states, and fosters neo-colonial posturing embedded in the instrumentation of drones for statecraft. As drones proliferate unchecked across Africa's battlefields, it underscores the urgent need for African leaders to grasp the complex nature of this development and formulate a nuanced strategy that addresses the long-term implications for the region's political stability and security landscape.

F. Okpaleke (✉)
Department of Political Science, University of Waikato, Hamilton, New Zealand
e-mail: okpalekefrancis@gmail.com

S. Oyewole
Federal University Oye-Ekiti, Oye, Ekiti, Nigeria

W. S. Coetzee and D. Putter (eds.), *Drones in the African Battlespaces*, Southern Space Studies, https://doi.org/10.1007/978-3-031-87147-4_3

1 Introduction

Over the past decade, Africa has seen a significant uptick in drone proliferation, with states integrating this technology to combat terrorism across regions like the Sahel and the Horn of Africa. Initially used in limited capacities, drones have become increasingly prevalent as tools for counterterrorism and addressing security challenges exacerbated by violent extremism and armed banditry.[1] The fragile security state is evident in the Security Threats Index, with ten of the twenty African countries ranking in the top 20 for terrorism. Mali holds the highest index score of 8.57.[2] This aligns with the Africa Security Threat (AST) index, which analyses security risks such as bombings, attacks, battle-related deaths, rebel activities, mutinies, coups, and terrorism. Data show that African nations had an index score triple that of European countries from 2007 to 2023, highlighting a persistent state of insecurity in the region.[3]

Amid these conflicts, drones are increasingly employed by African and foreign actors like the United States, France, Turkey, China, and Iran for counterterrorism and strategic purposes. This surge coincides with the strategic establishment of drone bases across Africa, reflecting an emerging trend termed the "dronification of the African battlespace".[4,5] Presently, more than 20 countries in Africa, including Ethiopia, Egypt, Nigeria, Togo, Angola, Chad, Morocco, South Africa, Niger, Burkina Faso, and Mali, possess drone technology.[6] Of note is the acquisition of armed drones by states previously lacking robust military arsenals, like Seychelles and Madagascar.[7] Concurrently, foreign exports to Africa have diversified, with China, Turkey, France, and Iran joining traditional suppliers like the US and Israel.[8] This expansion is complemented by establishing numerous drone bases across Africa, marking a transition from a platform-centric to a system-of-systems-centric business model. Previously dominated by the United States, several external actors, including France, India, Turkey, the United Arab Emirates,

---

[1] Okpaleke, F.N., Nwosu, B.U., Okoli, C.R. and Olumba, E.E., "The Case for Drones in Counter-Insurgency Operations in West African Sahel", *African Security Review* 32, no. 4 (2023), 351–367.
[2] Global Economy, "Security Threats Index—Country Rankings", 2023, www.theglobaleconomy.com/rankings/security_threats_index/Africa.
[3] Ibid.
[4] Kindervater, K.H., "Drone Strikes, Ephemeral Sovereignty, and Changing Conceptions of Territory", *Territorial Designs and International Politics* 119, no. 21 (2018), 93–107.
[5] Akhter, M., "The Proliferation of Peripheries: Militarized Drones and the Reconfiguration of Global Space", *Progress in Human Geography* 43, no. 1 (2019), 64–80. https://doi.org/10.1177/0309132517735697.
[6] Dass, R., "The Evolving Threat from Terrorist Drones in Africa", *LawFare*, 1 May 2023, www.lawfaremedia.org/article/the-evolving-threat-from-terrorist-drones-in-africa.
[7] Lionel, E., "US Plans Drone Bases in West Africa to Counter China and Islamist Militants", *Military Africa*, 15 January 2024, www.military.africa/2024/01/us-plans-drone-bases-in-west-africa-to-counter-china-and-islamist-militants.
[8] Dass, R., "The Evolving Threat from Terrorist Drones in Africa", 1 May 2023.

Japan, China, and Saudi Arabia, now have military bases operating in Africa to support drone operations.[9]

The proliferation of armed drones and drone bases across Africa manifests as a paradigm shift in the continent's drone market—from a theatre for drone attacks to becoming a big market for the export and power play by international actors.[10] Scholars suggest that Africa is rapidly becoming a focal point in a changing geopolitical landscape for drone warfare, with external actors strategically seeking to expand their regional influence through military and diplomatic means.[11,12,13] This is rooted in using drones as a tool for conducting diplomacy through other means—driven by Africa's strategic significance and existing security relationships.[14] Other scholars conceptualise this phenomenon as "drone colonialism," explaining how the export of drones serves as a tool of statecraft, allowing these actors to enhance their political influence and broaden their sphere of control within Africa.[15] Countries such as China and Turkey strategically deploy drones to the continent with intentions resembling Africa's colonial past—aiming to establish or reinforce a position providing security to safeguard their present and future national interests.[16] These technology-driven diplomatic efforts represent a dual process that influences both markets and geopolitics, leading to a shift toward countries offering the most affordable, user-friendly, and accessible drone platforms and systems with lax conditionalities.[17]

The current trend of drone proliferation in Africa also raises significant concerns about the potential for violent non-state actors (VNSAs), such as Al-Shabaab

---

[9] Horowitz, M., Schwartz, J.A. and Fuhrmann, M., "Who's Prone to Drone? A Global Time-Series Analysis of Armed Uninhabited Aerial Vehicle Proliferation", *Conflict Management and Peace Science* 39, no. 2 (2022), 119–142.

[10] Tankel, S., "US Counterterrorism in the Sahel: From Indirect to Direct Intervention", *International Affairs* 96, no. 4 (2020), 875–893.

[11] Paulsson, H., "Explaining the Proliferation of China's Drones", The Diplomat, 10 November 2018, https://thediplomat.com/2018/11/explaining-the-proliferation-of-chinas-drones.

[12] Weinberger, S., "China Has Already Won the Drone Wars", *Foreign Policy*, 10 May 2018, https://foreignpolicy.com/2018/05/10/china-trump-middle-east-drone-wars/.

[13] Chávez, K. and Swed, O., "The Proliferation of Drones to Violent Nonstate Actors", *Defence Studies* 21, no. 1 (2021), 1–24.

[14] Devore, M.R., "No End of a Lesson: Observations from the First High-Intensity Drone War", *Defense & Security Analysis* 39, no. 2 (2023), 263–266. https://doi.org/10.1080/14751798.2023.2178571.

[15] Allinson, J., "The Necropolitics of Drones", *International Political Sociology* 11, no. 9 (2015), 113–127. https://doi.org/10.1111/ips.12086 and Olumba, E. E. (2024). The necropolitics of drone bases and use in the African context. Critical Studies on Terrorism, 1–23.

[16] Eslami, M., "Iran's Drone Supply to Russia and Changing Dynamics of the Ukraine War", *Journal for Peace and Nuclear Disarmament* 5, no. 2 (2022), 507–518. https://doi.org/10.1080/25751654.2022.2149077; Watterson, P.J. and Sigman, R., "A Changing Landscape: Turkey's Drone Diplomacy in Sub-Saharan Africa", 2023, https://apps.dtic.mil/sti/trecms/pdf/AD1213030.pdf .

[17] Balbon, N., "Beyond Agent vs. Instrument: The Neo-Coloniality of Drones in Contemporary Warfare", *E-International Relations*, 3 August 2022, www.eir.info/2022/08/03/beyond-agent-vs-instrument-the-neo-coloniality-of-drones-in-contemporary-warfare.

and Boko Haram, to acquire cheap and accessible drones for premeditated attacks or weaponising commercial drones for lethal purposes.[18] The absence of robust international regulations to control drone sales and dissemination, coupled with the affordability and accessibility of commercial off-the-shelf drones, exacerbates this concern as evidenced by recent attacks by the Iran-backed Houthi faction on United States Naval vessels in the Red Sea.[19]

The rest of this chapter explores the literature on proliferation and the security dilemma theory of drone colonialism. This will facilitate the analysis, at the hands of neo-colonialism and realist theories and realist theory, of the geopolitics of drone proliferation in Africa and its security implications. The analysis further investigates drone acquisition, base establishment, and the implications of drone colonialism in African battlegrounds, highlighting the broader impact on non-state actors acquiring drones. This is followed by an analysis of United States drone bases in Africa, their operations, and their geopolitical proliferation nexus. The discussion then explores how 'triple frontier actors'—Iran, China, and Turkey, are reshaping the African drone market with the influx of drones into African battlespaces and the implication for indigenous drone manufacturing. The chapter conclude by rehashing the urgency for African leaders to develop regulations that protect Africa from the neo-colonial trappings of the emerging drone proliferation syndrome due to the long-term implications for the region's political stability and security.

2 Triadic Intersection: Security Dilemma, Proliferation and Drone Colonialism in Africa

Since 9/11, an increasing number of states have recognised the strategic value and tactical efficacy of drones in conflict zones, leading to a steady expansion in their acquisition and development. According to a report by the Stockholm International Peace Research Institute (SIPRI), over 60 out of the 197 countries currently possess drones, with this number expected to rise in the coming years as advancements in drone technology continue.[20]

At the conceptual level of analysis, the term "proliferation" takes on a nuanced significance when applied to drone technology, encompassing aspects of spread, diffusion, and expansion. While this superficially aligns with the basic understanding of the concept, scholars explored the embedded meaning of drone proliferation within the African context or global peripheries—formerly colonised

[18] Okpaleke, F.N., Drones and US Grand Strategy in the Contemporary World (Cham: Springer Nature, 2023).

[19] Grinberg, A., "Iran's Drone Diplomacy Is a Danger to the West", *Fathom Journal*, 17 April 2022, https://fathomjournal.org/irans-drone-diplomacy-is-a-danger-to-the-west/.

[20] SIPRI, "Arms Transfer Database", Stockholm International Peace Research Institute, 2020, www.sipri.org/databases/arms-transfers/sources-and-methods.

states. Akhter[21] interprets proliferation to occur at different levels. Akhter's conceptualisation divides proliferation into two distinct categories. Firstly, there is proliferation stemming from technologically advanced states to those at a technological disadvantage, such as from the United States to African countries positioned at the global periphery in terms of technological development. Secondly, there is what Chavez and Swed,[22] termed as "scattered peripheries," illustrating the widespread distribution of drones across different scales, often without regulation. This aligns with the notion of unchecked and unregulated drone proliferation. The former type of proliferation is governed by regulations, like those outlined in the Missile Technology Control Regime (MTCR), aimed at preventing sales and exports of drone technology to despots, rebel factions, and adversaries of the United States. Conversely, the latter type adopts a more laissez-faire approach, allowing drones to be bought and sold based on factors such as affordability, availability, and accessibility.[23]

Expanding on Akhter's concept of proliferation, the global landscape can be divided into two distinct spheres. Firstly, when drones proliferate into economically and militarily disadvantaged regions, it reshapes their understanding of territorial sovereignty, which becomes increasingly contingent within these states.[24] Secondly, the heightened deployment of drones in these internal peripheries creates sub-peripheries within them. This, in turn, gives rise to another phenomenon: neo-colonial spaces, comprising economically and militarily weaker regions in the global periphery dominated by more powerful states and internal peripheries within states deploying militarised drones.[25] These internal peripheries often serve as sites of resistance against perceived colonial dominance or as currently witnessed, resistance against the prevailing political incumbency. Akhter[26] emphasises the inherent asymmetrical power dynamics in drone proliferation, which extends to the establishment of bases in Africa. This dissemination of drone technology not only captures the underbellies of colonial character but also strengthens existing hierarchies of power and control, with dominant states exporting their military capabilities to economically and militarily weaker regions.[27] Additionally, the establishment of drone bases and their deployment within states'

[21] Akhter, "The Proliferation of Peripheries: Militarized Drones and the Reconfiguration of Global Space", 2019.

[22] Chávez and Swed, "The Proliferation of Drones to Violent Nonstate Actors", 2021.

[23] Weinberger, "China Has Already Won the Drone Wars", 10 May 2018 and Okpaleke et al., "The Case for Drones in Counter-Insurgency Operations in West African Sahel", 2023.

[24] Akhter, "The Proliferation of Peripheries: Militarized Drones and the Reconfiguration of Global Space", 2019.

[25] Chávez and Swed, "The Proliferation of Drones to Violent Nonstate Actors", 2021.

[26] Akhter, "The Proliferation of Peripheries: Militarized Drones and the Reconfiguration of Global Space", 2019.

[27] Espinoza, M., "State Terrorism: Orientalism and the Drone Programme", *Critical Studies on Terrorism* 11, no. 2 (2018), 376–393. https://doi.org/10.1080/17539153.2018.1456725 and Allinson, "The Necropolitics of Drones", 2015.

territories create internal peripheries, extending the concept of colonial control beyond conventional borders.

The above assertion is grounded in the theoretical framework of neo-colonialism, which suggests that powerful nations utilise drones as tools to maintain neo-colonial power structures. This perspective is echoed in the literature linking drone technology to neo-colonial forms of necropolitics.[28] From this viewpoint, drones serve as instruments of neo-colonialism wielded by the Global North to assert necropower by perpetuating insecurity in peripheral regions.[29] Necropolitics, in turn, reveal themselves through the designation of colonised populations as spaces of insecurity while providing security to colonisers.[30] Drones facilitate the creation of neo-colonial spaces, where marginalised groups face surveillance and state violence.[31] These spaces are constructed using sophisticated geospatial and geo-visualization technology, inherently culturally mediated, and exclusively populated by targets, enemies, and cultural others.[32] While acknowledging the cultural and techno-visual implications of drones, other scholars emphasise the colonial lineage of militarised drones through historical connections between capitalism and state violence—where drones function as an instrument for maintaining social order and asserting control. From this lens, drones are used as key military technologies for shaping global peripheries subordination to the security interests of "core countries", particularly Western nation-states or generally the selling state with strategic interests beyond the actual drone sales.[33]

While acknowledging the transformative nature of drones as military assets, which states inevitably seek to acquire due to their growing prominence in warfare, it is crucial to recognise their broader geopolitical significance. For instance, Turkish President Recep Tayyip Erdogan emphasised the importance of drone discussions during his 2021 tour of Africa, highlighting the emerging trend of utilising drones as tools for diplomacy and extending spheres of influence.[34] This approach is not unique to Turkey; other countries like China and Iran have also adopted similar strategies in Central Sahel and Ethiopia, respectively, further influencing both supply and demand dynamics from a market perspective, as well as determining winners and losers in the security landscape.[35] This aligns with the concept

[28] Espinoza, "State Terrorism: Orientalism and the Drone Programme", 2018.

[29] Balbon, "Beyond Agent vs. Instrument: The Neo-Coloniality of Drones in Contemporary Warfare", 3 August 2022.

[30] Mbembe, A., "Necropolitics", *Public Culture* 15, no. 1 (2003), 11–40.

[31] Akhter, "The Proliferation of Peripheries: Militarized Drones and the Reconfiguration of Global Space", 2019.

[32] Chávez and Swed, "The Proliferation of Drones to Violent Nonstate Actors", 2021.

[33] Watterson, "A Changing Landscape: Turkey's Drone Diplomacy in Sub-Saharan Africa", 2023.

[34] Bociaga, R., "Why Are Chinese Drones the Growing Choice in Africa?", *The Africa Report*, 10 April 2023, www.theafricareport.com/296467/why-are-chinese-drones-the-growing-choice-in-africa/.

[35] Haugstvedt, H., "A Flying Threat Coming to Sahel and East Africa? A Brief Review", *Journal of Strategic Security* 14, no. 1 (2020), 92–105, https://doi.org/10.5038/1944-0472.14.1.1848.

of drone colonialism, which elucidates how the proliferation of drone technology leads to colonial control and domination.[36] In this context, technologically advanced states with strategic interests leverage drones to assert influence and control over African airspaces, resources, and populations.[37]

A central feature of neo-drone colonialism lies in the inherent asymmetrical power dynamics within the spread of drone technology. Advanced states and corporations, armed with sophisticated drone capabilities, export their military technologies to economically and militarily weaker regions, reinforcing prevailing power and control hierarchies.[38] This diffusion of drone technology perpetuates colonial legacies of domination and exploitation, as external actors dictate the terms of engagement and exert control over African states.[39]

The increase in demand for drones worldwide highlights their emerging status as coveted assets, akin to weapon systems like nuclear weapons.[40] Historically, the United States has wielded significant influence in shaping the proliferation of drones in Africa, establishing drone bases and deploying drones against terrorist groups. This set the stage for what Kilcullen and Exum[41] construed as a 'death from above and outrage from below' to describe how the effectiveness of drones demonstrated by the United States inspired other states to embrace their technical and strategic utility in battlefield scenarios. The United States deployment of drones in targeted states following 9/11 served as a reference for its capability in warfare, demonstrating the efficacy of drone technology in remote warfare and providing support to ground troops while minimising casualties.

Initially championed by the United States for counterterrorism operations, the widespread use of drones has led other states to perceive a disparity in military-technological capabilities. In response, these states have begun acquiring their drone technology, or they recognise the potential application of these platforms within their national security frameworks. Horowitz et al.[42] construe the emerging "dronification" as both inevitable and militarily imperative. Despite the discourse surrounding the reality of dronification in Africa, evidence suggests a tangible

---

[36] Espinoza, "State Terrorism: Orientalism and the Drone Programme", 2018.

[37] Anyadike, O., "A Rough Guide to Foreign Military Bases in Africa", *The New Humanitarian*, 15 February 2017, www.thenewhumanitarian.org/feature/2017/02/15/updated-rough-guide-foreignmilitary-bases-africa and Balbon, "Beyond Agent vs. Instrument: The Neo-Coloniality of Drones in Contemporary Warfare", 3 August 2022.

[38] Cook, L., "Destabilizing Africa: Iranian Drones Menace Morocco", *The Defence Post*, 27 June 2023, www.thedefensepost.com/2023/06/27/iran-drones-morocco.

[39] Anyadike, "A Rough Guide to Foreign Military Bases in Africa", *The New Humanitarian*, 15 February 2017.

[40] Donnenfeld, Z., "Drone Strikes a Growing Threat to African Civilians", *Institute for Security Studies*, 27 February 2019, https://issafrica.org/iss-today/drone-strikes-a-growing-threatto-african-civilians.

[41] Kilcullen, D. and Exum, A.M., "Death from Above, Outrage Down Below", *New York Times*, 16 (2009), 529–535.

[42] Horowitz, M.C., Kreps, S. E. and Fuhrmann, M., "Separating Fact from Fiction in the Debate over Drone Proliferation", *International Security* 41, no. 2 (2016), 7–42.

shift underway regarding the role of drones as both a tactical and strategic tool for achieving disparate ends. However, one of the unintended effects of this dronification is the emergence of a security dilemma due to the resulting security spiral drones portend for Africa. The emergence of new technologies sparks a competitive rush among state actors seeking military advantage or the prestige of enhanced capabilities. Yet, one of the risks of this unregulated proliferation is the technology falling into the wrong hands, such as non-state actors, or causing other states to arm in response to the acquisition of drones by neighbouring states—leading to adverse regional and domestic security ramifications.

As conceptualised by John Herz, the security dilemma is characterised by six fundamental elements.[43] Firstly, it arises from the anarchic nature of the international system. Secondly, it stems from states' uncertainties and fears regarding each other's intentions. Thirdly, it prompts states to resort to self-help measures to ensure their security. Fourthly, it perpetuates a cycle of power competition among states. Fifthly, it highlights that the mere accumulation of power does not always equate to increased security. Lastly, it operates as a self-reinforcing and vicious cycle. Within the context of the security dilemma, the state is the primary object of security, viewed primarily through a military lens. While this theoretical framework is traditionally applied to international relations, where states are perpetually uncertain about one another's intentions in a realist world,[44] it can also be aptly appropriated to explain the current proliferation of drones as a manifestation of the security dilemma in Africa. It is worth noting that the use of the security dilemma theory to explain drone proliferation in Africa highlights the dual tensions within the continent's geopolitical landscape. Meanwhile, neo-colonial theory provides insights into the historical contexts that inform current patterns of drone supply and demand and the increasing militarization of African airspace. The security dilemma focuses on the risks posed by unchecked drone proliferation. While these theories have structural similarities, their application in this chapter is not to blur their distinct approaches but to clarify their relevance to the issue of drone proliferation in Africa.

Viewed through this lens, armed drones offer militaries distinct asymmetric advantages in battlefield scenarios, potentially signalling to other states an intent to bolster their armament capabilities. Procuring drones can thus communicate to other nations that a state is gaining strategic advantages, amplifying the uncertainty inherent in this dilemma.[45] For instance, in 2021, Morocco's acquisition of Chinese and Israeli drones, primarily to bolster its military and security operations against the Polisario Front in Western Sahara, prompted its northern neighbours, Algeria and Tunisia, to enhance their arsenals with Turkish and Chinese

[43] Herz, J.H., "The Security Dilemma in International Relations: Background and Present Problems", *International Relations* 17, no. 4 (2003), 411–416.

[44] Kaufmann, C. (2007). A security dilemma. *Harvard International Review* 28(4), 44–49.

[45] Tang, S., "The Security Dilemma: A Conceptual Analysis", *Security Studies* 18, no. 3, (2009) 587–623.

drones. States, driven by defence imperatives, often amass offensive capabilities defensively, contributing to a cycle of power accumulation.[46]

The proliferation of drones among global powers, including the United States and France, has significantly contributed to an emerging security dilemma. This dilemma is exemplified by the use of drones in counterterrorism missions across Africa, as seen in countries like Mali. Furthermore, other states have joined in exploiting loopholes in international agreements, such as the MTCR to export drones to Africa. Favourably, China, Iran, and Turkey are not among the 42 participating states of the Wassenaar Arrangement, which collectively regulates the export of munitions and dual-use goods and technologies.[47]

The continued proliferation of drones, coupled with the security dilemma, significantly increases vulnerabilities within Africa's already fragile security architecture, as violent non-state actors (VNSAs) are likely to incorporate drones into their arsenals by shown by recent events. For instance, there have been over 170 attacks on United States bases in Syria and Iraq between October 2023 and February 2024, resulting in nearly 200 injuries and three fatalities among US Forces troops, following non-state armed group responses to Israeli actions in Gaza. Similarly, the Houthi rebels in Yemen backed by Iran, have targeted international shipping in the Red Sea with modified drones and missiles for months.[48] In sub-Saharan Africa, various groups like Al-Qaeda in the Islamic Maghreb (AQIM), Ansar Dine, Movement for Oneness and Jihad in West Africa (MOJWA), and Islamic State in the Greater Sahara (ISGS) have reportedly employed hobbyist or modified drones in their operations across regions including Nigeria, Chad, Mali, and Niger. Groups such as Boko Haram have utilized hobbyist drones for surveillance and attacks against state forces.[49] This pattern of dronification among non-state actors such as Boko Haram, which uses commercial drones for lethal assaults on Nigerian military targets—is indicative of how the use of drones by other non-state actors at the international and regional level can serve to embolden others non-state actors to acquire, modify and use drone technology for a range of purposes including propaganda, plan attacks and for power projection.[50] Without stringent international regulations to control drone sales and distribution, and given the affordability and accessibility of commercial drones, there is a growing risk that non-state could increasingly acquire and deploy military-grade drone technology, as seen in the recent drone assaults by Iran-backed Houthi forces on

[46] Ibid.
[47] Kreuzer, M.P., "The Context and Prospects of Regulating Drones in Conflict", Drones and Global Order (2021), 209–226.
[48] Gambrell, J., "Houthis vow to retaliate after US, British strikes; Bombing of multiple sites in Yemen is in response to drone and missile attacks on ships in the Red Sea", *Globe & Mail*, 13 January 2024.
[49] Okpaleke et al., "The Case for Drones in Counter-Insurgency Operations in West African Sahel", 2023.
[50] Ibid.

United States naval vessels in the Red Sea to perpetrate attacks in the future.[51] For state actors the desire of neighbouring states to acquire drones, like other military technologies, signals to other states of their intentions and increases the propensity for other states to acquire drones, thereby, heightens uncertainty about the intentions of other actors in a realist world.

3 Geopolitics of Drone in Africa: State Actors and Non-state Actors

The geopolitical landscape of drone proliferation in Africa depicts a complex network of exports, sales, acquisitions, and the establishment of drone bases, with profound implications for state power, sovereignty, security, and international relations—that paints Africa as the new battleground for the contestation among external actors vying for strategic control of the African market. Scholars argue that these sales extend beyond economic motives, representing strategic manoeuvres akin to a colonial playbook.[52,53,54] Viewed through this lens, drone sales not only reflect a neo-colonial character aimed at using them as instruments to assert control in peripheral countries but also as a means through which states are beginning to pursue expansionist agendas.[55] For example, while Niger has requested the United States to vacate Camp 201 in Agadez, its primary drone base, it has welcomed Russian and Iranian military personnel to enter the airbase previously occupied by US troops, which violates Niamey's status of forces agreement with Washington.[56] This shift in alliances follows the coup by the junta in Niger, prompting the United States to withdraw its nearly 1000 military personnel from the country, which had been a key partner for Washington's counterinsurgency efforts until the coup occurred last year.[57] Also, despite tensions between Iran and France and the United States, France's withdrawal from the Sahel and its limitation on drone usage for aerial warfare prompted Mali, Burkina Faso, and Niger to increase their stock of Turkish-made drones.[58] Paradoxically, although France withdrew its military presence from the Sahel following the end of Operation Barkhane, it continued to maintain its neo-colonial influence in the Sahel. These include an air force base in N'Djamena, Chad, a regional base in Gao,

---

[51] Ibid.

[52] Mbembe, "Necropolitics", 2003.

[53] Grinberg, "Iran's Drone Diplomacy Is a Danger to the West", *Fathom Journal*, 17 April 2022.

[54] Cannon, "Out of Africa: The Impact of Drones in Sub-Saharan Conflicts", 2023.

[55] Jacobsen, M., "The Strategic Implications of Non-State Warbots", *The Strategy Bridge*, 15 December 2017, https://thestrategybridge.org/the-bridge/2017/12/15/the-strategic-implicationsof-non-state-warbots.

[56] Bociaga, "Why Are Chinese Drones the Growing Choice in Africa?", 10 April 2023.

[57] Lin-Greenberg, E, "The dawn of drone diplomacy", *Foreign Affairs* (20 December 2022), https://www.foreignaffairs.com/ukraine/dawn-drone-diplomacy and https://www.cnas.org/publications/commentary/the-dawn-of-drone-diplomacy.

[58] Watterson, "A Changing Landscape: Turkey's Drone Diplomacy in Sub-Saharan Africa", 2023.

Mali, a special forces base in Ouagadougou, Burkina Faso, and an intelligence base in Niamey, Niger, which coordinates drone operations and intelligence gathering across the Sahel.[59]

The geopolitics of drone proliferation in Africa transcends mere commercial transactions, embodying a complex interplay of strategic interests, regional dynamics, and the legacy of colonial power structures, each contributing to the widespread adoption of this technology. Firstly, cost plays a significant role. Chinese drones offer a cost-effective alternative to their United States counterparts, making them accessible to countries with limited defence budgets.[60] For instance, "the CH-4 and the Wing Loong 2 Chinese drones are estimated to cost between $1 m and $2 m, while the United States Reaper drones cost $16m and the Predator $4 m".[61] The cheaper price tag means interested governments can also purchase the drones in larger quantities. Irrefutably, Chinese drones offer African countries an appealing alternative to United States models due to their lower cost. However, this affordability often compromises quality and capabilities, especially concerning loiter time and range. The significant emergence of China as a supplier of militarised drone technology underscores the United States's hesitancy to export to certain states, revealing broader power dynamics.[62]

Secondly, the hyper-competitive nature of the international arms market fuels drone proliferation. China's emergence as a significant supplier challenges the dominance of traditional exporters like the United States.[63] This competition fosters innovation and diversification in drone offerings, providing states with a broader array of options to choose from. These drones are already reshaping the nature of warfare in Africa. Apart from United States drones, the variety of drone offerings in Africa includes those from Russia, Iran, and Turkey, with China leading the pack in production and adoption across North Africa, the Sahel, and the Horn of Africa. For example, in Sudan, the Sudanese Armed Forces and fighters with the paramilitary Rapid Support Forces (RSF) have been emboldened by the use of Iranian-made drones.[64] In July 2023, an RSF drone attack on a hospital in Omdurman resulted in numerous civilian casualties, with the Iranian-made Mohajer-6 drone being linked to the attack.[65] The Sudan War Monitor warns that

---

[59] Lionel, "US Plans Drone Bases in West Africa to Counter China and Islamist Militants", 15 January 2024.

[60] Bociaga, "Why Are Chinese Drones the Growing Choice in Africa?", 10 April 2023.

[61] Rasheed, Z., "How China became the world's leading exporter of combat drones", *Al Jazeera*, 24 January 2023, https://www.aljazeera.com/news/2023/1/24/how-china-became-the-worlds-leading-exporter-of-combat-drones.

[62] Moore, A. and Walker, J., "Tracing the US Military's Presence in Africa", *Geopolitics* 21, no. 3 (2016), 686–716.

[63] Amberg, P, Kim, E; Shin R, Godfray J, Dive E, and Devgan E., "The US and UK Announce Further Sanctions and Export Restrictions Targeting Iran", *Sanctions News*, 2 May 2024, https://sanctionsnews.bakermckenzie.com/the-us-and-uk-announce-further-sanctions-and-export-restrictions-targeting-iran/.

[64] Devore, "No End of a Lesson: Observations from the First High-Intensity Drone War", 2023.

[65] Cook, "Destabilizing Africa: Iranian Drones Menace Morocco", 27 June 2023.

Iran's intervention on the side of the SAF poses a risk of escalating Sudan's internal conflict into a proxy war between Iran and the United Arab Emirates (UAE), further complicating the dynamics of the conflict. Cook contends that Iran is employing drones as a destabilising tool in Africa, seeking to establish a presence in strategically important and resource-rich regions near vital Western shipping lanes in the Atlantic Ocean.[66] This strategy unfolds amidst Iran's broader expansion and consolidation of influence in the Sahel and Maghreb regions. Iran's tactics mirror its approach in the Middle East, as evidenced by its support for rebel Shiite groups through arms provision and training in countries such as Senegal, the Central African Republic, and Nigeria.[67]

Espinoza contextualises the burgeoning drone exports in Africa within the evolving power dynamics of the global inter-state system.[68] While the United States seeks to restrict the proliferation of armed drones to its allies, other nations are capitalising on the opportunity by increasing their drone sales to various countries, including those considered adversaries by the United States.[69] Countries like Turkey, Iran, and, notably, China have recently expanded their exports to African states. This trend positions Africa as a focal point for two distinct phenomena. Firstly, it becomes a new battleground for drone exportation, where counterterrorism missions serve as gateways to penetrate the African market and flood it with drones that lack regulatory barriers and are more cost-effective compared to those from the United States and Israel.[70] Secondly, overlooking the inherent geopolitics at play neglects the zero-sum dynamics evident in how former colonial powers still exert influence over their former colonies. France, for example, maintains significant military and economic leverage over its former colonies, despite recent challenges such as coups in Mali, Niger, and Gambia, as well as France's withdrawal from the region. However, this does not erase the enduring influence these colonial legacies wield, shaping both the supply and demand dynamics in the geopolitical landscape.[71] Geo-supply reflects the ability of former colonies to engage in bilateral trade and establish alliances. This dynamic is deeply rooted in the center-periphery theory of dependence, where aid, strategic partnerships, and agreements serve as tacit tools for exerting control over these countries. On the geo-demand side, the need for military assistance to support troops creates a

---

[66] Ibid.

[67] Weinberger, "China Has Already Won the Drone Wars", 10 May 2018.

[68] Espinoza, "State Terrorism: Orientalism and the Drone Programme", 2018.

[69] Horowitz, et al., "Separating Fact from Fiction in the Debate over Drone Proliferation, 2016.

[70] Schwartz, J.A., Fuhrmann, M. and Horowitz, M.C., "Do Armed Drones Counter-Terrorism, or Are They Counterproductive? Evidence from Eighteen Countries", *International Studies Quarterly* 66, no. 3 (2022), 42–47.

[71] Balbon, "Beyond Agent vs. Instrument: The Neo-Coloniality of Drones in Contemporary Warfare", 3 August 2022.

dependency relationship reminiscent of the adage "he who pays the piper calls the tune".[72]

Lin-Greenberg[73] and Watterson[74] frame the surge in drone sales by China, Iran, and Turkey as a form of drone diplomacy, wherein these foreign actors exploit gaps in arms control regimes to bolster their global influence in Africa. Through drone sales, these states wield drones as diplomatic instruments to advance their foreign policy agendas, extending beyond mere economic transactions. This encompasses activities such as donating or selling drone weaponry to secure concessions, expand influence, counter rivals, and strengthen military ties.[75] Furthermore, states such as China, Japan, India, and Russia (through entities like the Wagner Group) possess the capability to assert dominance over the skies of African countries using killer drones or surveillance drones. These state actors may pursue their strategic interests, including resource extraction, geopolitical influence, or counterterrorism efforts by establishing drone bases.[76]

4 United States and Drone Bases in Africa

The United States remains a major player in the deployment of drones with established presence and operations in Africa, making any discussion about drones on the continent incomplete without considering their role and influence. While other Western nations, such as France and the United Kingdom, have also deployed drones, their presence does not match the scale or impact of the United States.[77] The scope of United States drone operations in Africa is extensive, with approximately 27 bases dispersed across 15 African countries.[78] Notably, in Somalia, the United States maintains a base at Baledogle Airfield, utilised for drone operations and other military functions. Additionally, the Joint Special Operation Command (JSOC), which is rumoured to be located in Kismayo, Somalia, operates a secretive base, contributing to counterinsurgency efforts against groups like al-Shabab.[79] Somali forces rely on United States intelligence support as they combat al-Shabab

[72] Okpaleke, et al., "The Case for Drones in Counter-Insurgency Operations in West African Sahel", 2023.
[73] Lin-Greenberg, "The dawn of drone diplomacy", *Foreign Affairs* (20 December 2022).
[74] Watterson, "A Changing Landscape: Turkey's Drone Diplomacy in Sub-Saharan Africa", 2023.
[75] Lin-Greenberg, E, "The dawn of drone diplomacy", *Foreign Affairs* (20 December 2022.
[76] Lushenko, P. and Kreps, S., "Americans Support Exporting Drones to Ukraine — With a Caveat", *Brookings*, 2023, www.brookings.edu/articles/americans-support-exporting-drones-to-ukraine-with-a-caveat.
[77] Abunahel, M., "The United States Has Built a Network of Drone Bases Across Africa", *World Beyond War*, 22 March 2024, https://worldbeyondwar.org/the-united-states-has-built-a-network-of-drone-bases-across-africa/.
[78] Rogers, J. and Goxho, D., "Light Footprint—Heavy Destabilizing Impact in Niger: Why the Western Understanding of Remote Warfare Needs to Be Reconsidered", *International Politics* 60, no. 4 (2023), 790–817. https://doi.org/10.1057/s41311-021-00362-9.
[79] Moore and Walker, "Tracing the US Military's Presence in Africa", 2016.

and other militant groups, bolstered by a combination of AU-led counterinsurgency efforts, airstrikes, and United States special operations raids.[80] Furthermore, on February 15, 2024, the United States and Somalia inked a Memorandum of Understanding (MoU) for the construction of up to five military bases for the Somalia National Army's Danab Brigade, with an estimated value of $100 m.[81] Moreover, the United States has established a presence in Ethiopia, a key ally in the Horn of Africa. Moreover, the United States has strategically positioned drone bases in vital locations such as Djibouti, where they conduct counterterrorism operations spanning the Horn of Africa and the Arabian Peninsula. Additionally, drone bases have been established in the Republic of Seychelles to facilitate counterpiracy surveillance, further extending the United States' regional operational reach.[82] Despite this widespread presence, the United States also leverages host nation facilities in nations like Senegal and Uganda, demonstrating its capacity to operate within the region even without formal base agreements.

Furthermore, the expansion of U.S. drone operations in Africa appears inexorable, as the U.S. strategically extends its drone bases into East Africa and establishes a presence in Coastal West Africa, including countries such as Ghana, Benin Republic, Mali, and Ivory Coast, aiming to establish unarmed reconnaissance airfields as part of its efforts to combat the spread of al-Qaeda and Islamic State in the region and counter the dual threats of China and Islamist militants in West Africa.[83,84] Additionally, the U.S. aims to deter and disrupt China's plans to establish its first permanent naval base on the Atlantic coast in Equatorial Guinea, which poses a direct challenge to U.S. national security and interests, as highlighted in United States intelligence reports revealing that China's naval warships would be able to operate opposite the East Coast of the U.S. China has been expanding its economic and military presence in Africa, offering loans, infrastructure projects, and arms sales to various countries. This development signals a bolstering of regional drone capabilities for the U.S.[85]

In contrast to the significant presence of external actors like the United States and France, no African country currently hosts a drone base or actively pursues the acquisition of one. This power asymmetry raises concerns about the unchecked authority of foreign powers to conduct surveillance and targeted killings within African nations, effectively assuming roles akin to judges, juries, and executioners

---

[80] McCormick, T.Y., "Exclusive: U.S. Operates Drones from Secret Bases in Somalia", *Foreign Policy*, 2 July 2015, https://foreignpolicy.com/2015/07/02/exclusive-u-s-operates-drones-from-secret-bases-in-somalia-special-operations-jsoc-black-hawk-down/.

[81] Lionel, "US Plans Drone Bases in West Africa to Counter China and Islamist Militants", 15 January 2024.

[82] Donnenfeld, "Drone Strikes a Growing Threat to African Civilians", 27 February 2019.

[83] Ajala, O., "US Drone Base in Agadez: A Security Threat to Niger?", *RUSI Journal* 163, no. 5 (2018), 20–27. https://doi.org/10.1080/03071847.2018.1552452.

[84] Cannon, "Out of Africa: The Impact of Drones in Sub-Saharan Conflicts", 2023.

[85] Rogers and Goxho, "Light Footprint—Heavy Destabilizing Impact in Niger: Why the Western Understanding of Remote Warfare Needs to Be Reconsidered", 2023.

in the region. The presence of U.S. drone bases in Africa has not only drawn the involvement of its allies in international operations, such as the United Nations Stabilization Mission in Africa, but has also led to collaborative efforts in counterterrorism. For instance, countries like Italy have deployed drones to Libya under *Operation Atalanta* from the U.S. Chabelley Air Base in Djibouti, while Sweden and the Netherlands have utilised US drone bases for similar purposes.[86] While this highlights international cooperation, it carries strong connotations of drone colonialism, as the establishment and operation of these drones facilitate external actors' operations within African territories.[87]

By deploying drones for surveillance and control, external actors extend their influence into African airspace and territories, solidifying their dominance. This nexus between drones and capitalism underscores the economic motives behind neo-drone colonialism—which has been construed as the continuation of colonialism through other means.[88] In this sense, drones serve the interests of capitalist states and corporations by enabling resource exploitation, suppression of dissent, and maintenance of social order. This economic exploitation perpetuates colonial legacies of resource extraction and economic dominance, perpetuating dependence (military support and training) in African nations.[89]

5 Triple Frontiers: Turkey, Iran, and China and Drone Sales in Africa

Across the African continent, from the Sahel to the Horn of Africa, numerous states are amassing and deploying drones manufactured by China, Iran, and Turkey to address a spectrum of threats, spanning from terrorism to armed banditry. Countries such as Nigeria, Algeria, Egypt, Morocco, and Ethiopia have integrated drones from these countries into their military arsenals. According to the SIPRI, which monitors global arms transfers, China has exported armed drones to over 17 nations in the past decade, with approximately half of these sales directed to African countries.[90] Projections indicate that this trend will likely continue, with China poised to become one of the world's foremost exporters of armed drones. In contrast, the United States has supplied armed drones to only twelve countries

[86] Eslami, "Iran's Drone Supply to Russia and Changing Dynamics of the Ukraine War", 2022.
[87] Amberg et al., "The US and UK Announce Further Sanctions and Export Restrictions Targeting Iran", *Sanctions News*, 2 May 2024.
[88] Olumba, "The necropolitics of drone bases and use in the African context", *Critical Studies on Terrorism*, (2024), 1–23.
[89] SIPRI, "Arms Transfer Database", 2020.
[90] Lushenko and Kreps, "Americans Support Exporting Drones to Ukraine—With a Caveat", *Brookings*, 2023.

during the same period, primarily to allies such as France, the United Kingdom, the UAE, and most recently, Ukraine.[91]

The 'triple frontier' countries appear to have greater control of the African drone space. Ethiopia employs a diverse array of drones, including those from Chinese, Turkish, Emirati, and Iranian origins, for surveillance and combat purposes in civil conflicts in Tigray and Oromo.[92] Nigeria also utilises Chinese drones in its efforts against Boko Haram and is concurrently developing its own armed drones with support from Pakistan.[93] Cameroon acquired small military drones in 2018, and Kenya employs drones for monitoring al-Shabaab insurgents in neighbouring Somalia.[94] Beyond surveillance, countries like the Democratic Republic of the Congo and Mozambique have deployed drones in combat.[95] Additionally, Botswana, Côte d'Ivoire, Sudan, and Zambia have integrated mid-sized tactical drones into their military arsenals.[96] The success of Turkish Bayraktar TB2 drones in Libya's civil war has elevated the profile of weaponised drones, prompting acquisitions by militaries in Ethiopia, Morocco, Rwanda, Togo, and Tunisia, and potential consideration by Angola.[97]

Iran's drone export activities have expanded to encompass three continents: Europe, Africa, and Asia. Scholars argue that Iran targets countries experiencing ongoing armed conflicts as potential markets for its drone sales.[98] While the veracity of these claims remains subjective, it is undeniable that Iran faces sanctions from both the UK and the United States, with both countries imposing new sets of sanctions and export restrictions specifically targeting Iranian drones. Nevertheless, Africa presents a new frontier for Iran to market its drone technology, with its primary exports consisting of long-range Shahed-136 suicide drones and shorter-range Mohajer-6 attack drones.[99]

Currently, Iran supplies drones to the Sudanese government, which employs them against insurgent groups like the Sudanese People's Liberation Movement and others operating in Darfur. Moreover, in 2021, Iran sold drones to Ethiopia for use against Tigrayan rebels. Additionally, Iran has confirmed sales of its drones to

---

[91] Bashfield, S., "Agalega: A Glimpse of India's Remote Island Military Base", *The Interpreter*, 18 May 2021, www.lowyinstitute.org/the-interpreter/agalega-glimpse-india-s-remote-island-military-base.

[92] Eslami, "Iran's Drone Supply to Russia and Changing Dynamics of the Ukraine War", 2022.

[93] Haruna, A., "Drone Sighted Before Deadly Ambush on Security Forces in Northeast Nigeria", *The Human Angle*, 2022, https://humanglemedia.com/drone-sightedbefore-deadly-ambush-on-security-forces-in-northeast-nigeria/%0A.

[94] Amberg et al., "The US and UK Announce Further Sanctions and Export Restrictions Targeting Iran", *Sanctions News*, 2 May 2024.

[95] Lin-Greenberg, "The dawn of drone diplomacy", *Foreign Affairs* (20 December 2022.

[96] Lionel, "US Plans Drone Bases in West Africa to Counter China and Islamist Militants", 15 January 2024.

[97] Haugstvedt, "A Flying Threat Coming to Sahel and East Africa? A Brief Review", 2020.

[98] Grinberg, "Iran's Drone Diplomacy Is a Danger to the West", *Fathom Journal*, 17 April 2022.

[99] Gonçalves, M., "The Scramble for Africa Reloaded? Portugal, European Colonial Claims and the Distribution of Colonies in the 1930s", *Contemporary European History* 30 (2021), 2–15.

Algeria, a country with which it has longstanding economic and military ties.[100] Across the Red Sea in Yemen, Iran has provided drones to Houthi rebels, who utilise them in attacks against the UAE and Saudi Arabia. Scholars suggest that the proliferation of Iranian drones in Sudan could potentially trigger a security dilemma, prompting violent non-state actors and rebel groups in northern and central Africa to seek these drones for conducting attacks, thereby undermining regional security architecture.[101,102]

Turkey drone exports to Africa have surged over the past decade, with SIPRI reporting that Turkey currently supplies armed drones to about ten countries on the continent, spanning from Central Sahel to the Horn of Africa.[103] These include Nigeria, Morocco, Ethiopia, Mali, Togo, and Burkina Faso. For example, Niger employs these drones against insurgents operating in the Sahara Desert and Lake Chad, while Nigeria uses them against Boko Haram. Additionally, they supported the Libyan Government of National Unity in countering Marshal Khalifa Haftar's offensive and enabled Ethiopian Prime Minister Abiy Ahmed to outmanoeuvre Tigray rebels.[104] These drones, notably the Bayraktar TB2s, are cost-effective and high-performing, demonstrating their capabilities in conflicts such as those in Azerbaijan and Ukraine. SIPRI estimates show that exports, which were a mere $248 m in 2002, soared to $4.4 billion in 2022 and are projected to reach $10 billion in the next decade.[105] This expansion of Turkey's drones into Africa serves as a means to project its influence while serving domestic economic and political imperatives.[106] However, this development poses a potential threat to the interests and policies of the United States across the region, albeit still in its early stages.

While there is limited evidence suggesting that Turkey's emergence in the drone market will prompt the United States to reconsider its relations with specific countries, it is notable that Turkey has successfully obtained concessions from nations in exchange for its drone technology. For instance, Nigeria granted concessions in the mineral sector, and Ethiopia closed Gülenist schools as a result.[107] If Turkey continues to secure concessions or achieve outcomes that directly affect United States interests, such as enhancing regional stability or countering violent extremist groups, a clearer correlation between drone sales and U.S. interests may emerge.[108]

[100] Devore, "No End of a Lesson: Observations from the First High-Intensity Drone War", 2023.
[101] Schwartz et al., "Do Armed Drones Counter-Terrorism, or are they Counterproductive? Evidence from Eighteen Countries", 2022.
[102] Lushenko and Kreps, "Americans Support Exporting Drones to Ukraine—With a Caveat", *Brookings*, 2023.
[103] Watterson, "A Changing Landscape: Turkey's Drone Diplomacy in Sub-Saharan Africa", 2023.
[104] Eslami, "Iran's Drone Supply to Russia and Changing Dynamics of the Ukraine War", 2022.
[105] SIPRI, "Arms Transfer Database", 2020.
[106] Cook, "Destabilizing Africa: Iranian Drones Menace Morocco", 27 June 2023.
[107] Paulsson, "Explaining the Proliferation of China's Drones", 10 November 2018.
[108] Dass, R., "Militants and Drones: A Trend That is Here to Stay", *The Royal United Services Institute (RUSI)*, 6 September 2022, https://www.rusi.org/explore-our-research/publications/commentary/militants-and-drones-trend-here-stay.

China, in particular, has emerged as a dominant force in the global drone market, leveraging its robust manufacturing capabilities to cater to the needs of unstable and militarised regions, which have become a recurrent feature of the African continent. However, China is not the sole beneficiary of the burgeoning drone market in Africa; other foreign actors, including Iran and Turkey, are capitalising on the escalating tensions to penetrate and saturate the African market with drones.[109] However, akin to the Trojan Horse, scholars have pointed out the hidden agendas behind these sales, with motives ranging from colonial legacies to the perils of globalisation and Africa's positioning as a peripheral player on the global stage. Some argue that the proliferation of drones in Africa reflects new forms of diplomacy driven by technological advancements, market capture strategies, and lax arms control regimes, with drones serving as a tool for the instrumentation of statecraft through other means.[110] Nevertheless, there is no one-size-fits-all analysis; these various elements coalesce around the central theme of security disequilibrium, which alters the landscape, interpretation, and manifestation of how drones are diffusing throughout Africa.

The emerging tripartite frontier of external actors, each pursuing their strategic interests, is reshaping the agenda of drone proliferation in Africa, marking a new geopolitical landscape. It appears inevitable that this trend will proliferate, introducing security vulnerabilities and fragilities between the United States and the 'triple frontier' as they vie to maintain influence in Africa and control access to drone technology by states and non-state actors. This dynamic could potentially escalate conflicts across African regions, adding another dimension to the existing security challenges.

6 Indigenised Drone Manufacture in Africa

Indigenous drone production is emerging as a significant facet of Africa's engagement with drone technology, further enriching the discourse surrounding its proliferation. While African nations have long been recipients of drone imports, the rise of domestic drone manufacturing signifies a notable shift towards self-reliance and technological advancement. Currently, more than seven African countries are actively producing drones for various purposes, ranging from surveillance and reconnaissance to combat operations.[111] Although these domestically manufactured drones have yet to be exported beyond the continent, their production reflects the growing integration of drone technology into African military systems, portending potential implications for regional security dynamics.[112]

109 Schwartz et al., "Do Armed Drones Counter-Terrorism, or are they Counterproductive? Evidence from Eighteen Countries", 2022.

110 Watterson, "A Changing Landscape: Turkey's Drone Diplomacy in Sub-Saharan Africa", 2023.

111 SIPRI, "Arms Transfer Database", 2020.

112 Okpaleke, F. and Okoli, A.C., "Drones and Counter-Banditry Operations in the Western Sahel: Strategic Prospects", in *Banditry and Security Crisis in Nigeria* (2023), 210–221.

At the regional level, Southern Africa leads in the production of locally made drones, boasting 94 units across twelve models. North Africa closely follows with 51 units across 14 models, while West Africa and East Africa have produced 20 units each. Notable examples include Nigeria's Tsaigumi UAV, developed in collaboration with Portugal-based UAVision, and Kenya's reverse-engineered version of Israel's Aerostar drone, dubbed 'TAI'.[113] South Africa stands out as Africa's preeminent drone manufacturer, with its company Denel leading the production of fixed-wing, long-endurance drones tailored for missions such as search and rescue, maritime patrol, and precision airstrikes.[114]

Egypt has also made strides in drone production and manufacturing models like the EJune-30SW under the auspices of the Ministry of Military Production. Collaborative efforts with Chinese defence manufacturer Xi'an Aisheng Technology Group have also yielded models like the ASN-209 UAV.[115] Star Aviation has developed the Amel 300 drones in Algeria, further contributing to the continent's burgeoning indigenous drone capabilities.[116] The proliferation of domestically manufactured drones in Africa represents a significant step towards technological autonomy and self-sufficiency. However, it also raises questions about the potential consequences of unchecked drone proliferation, particularly concerning regional security dynamics and the risk of escalation in conflicts, particularly if non-state actors acquire lethal drone manufacturing capabilities.

7 Violent Non-state Actors and Drones

It is not just the state actors who are deploying drones in Africa; al-Shabaab, the Libyan Liberation Army, ISIS-Mozambique (Ansar al-Sunna), Boko Haram, and the Islamic State's West Africa Province (ISWAP) deploy drones for various tactical purposes, including intelligence, surveillance, and reconnaissance (ISR), as well as for producing propaganda videos.[117] These groups often challenge the authority of the state and engage in activities ranging from armed conflict to terrorism, insurgency, and organised crime. Several of these groups, including Al-Qaeda in the Islamic Maghreb (AQIM), Ansar Dine, Movement for Oneness and Jihad in West Africa (MOJWA), and Islamic State in the Greater Sahara (ISGS), operate

[113] Haruna, "Drone Sighted Before Deadly Ambush on Security Forces in Northeast Nigeria", 2022.

[114] Cannon, "Out of Africa: The Impact of Drones in Sub-Saharan Conflicts", 2023.

[115] Grinberg, "Iran's Drone Diplomacy Is a Danger to the West", *Fathom Journal*, 17 April 2022.

[116] Ibid.

[117] Derek, G., "Theory of the Drone: Historical Precedent and Postcolonial Amnesia", *Geographical Imaginations*, 8 August 2013, https://geographicalimaginations.com/2013/08/08/theory-of-the-drone-7-historical-precedentand-postcolonial-amnesia/; Haruna, "Drone Sighted Before Deadly Ambush on Security Forces in Northeast Nigeria", 2022.

in sub-Saharan Africa and are frequently categorised as terrorists or illegitimate entities.[118]

Armed groups like Boko Haram are actively weaponizing small commercial drones for surveillance and combat, heightening security risks for African governments. This surge in drone usage among insurgents intensifies security challenges across the continent. The potential for drones to shape future conflicts in Africa has garnered increased attention. Concerns are mounting over the possibility of terrorist groups like ISWAP, Al Shabaab, and Boko Haram weaponizing off-the-shelf drones or acquiring advanced armed drones.[119] The militarization of drone technology in Africa introduces an asymmetrical tool that violent non-state groups can exploit, fundamentally altering conflict dynamics on the continent.[120] While most hobbyist or commercial off-the-shelf drones used by these terrorists serve passive defensive purposes, such as conducting ISR before attacks and creating propaganda videos, they also symbolize airpower, status, and technological prowess, potentially aiding in recruitment and securing donations.[121]

8 Discussion

The proliferation of drones in Africa can be attributed to various factors, categorised as external, internal, and outliers. External factors, driven by foreign actors, significantly drive drone proliferation.[122] The escalation of terrorism and insurgent activities has prompted interventions from external powers, notably the United States, France and United Nations Multidimensional Integrated Stabilization Mission in Mali (MINUSMA) West Africa, employing drones in operations such as Takuba. Beyond external influences, outliers—intersecting internal and external motivations—also shape the current pattern of dronification in Africa. Scholars argue that dronification is a result of both security and innovation dilemmas associated with military technologies.[123,124] The emergence of new technologies triggers a scramble among state actors to acquire them, driven by the desire for military advantage or the status symbol of bolstering military capabilities. Additionally,

[118] Okpaleke et al., "The Case for Drones in Counter-Insurgency Operations in West African Sahel", 2023.

[119] Haugstvedt, "A Flying Threat Coming to Sahel and East Africa? A Brief Review", 2020.

[120] Lushenko and Kreps, "Americans Support Exporting Drones to Ukraine—With a Caveat", *Brookings*, 2023.

[121] Oyewole, S., "Making the Sky Relevant to Battle Strategy: Counterinsurgency and the Prospects of Air Power in Nigeria", *Studies in Conflict & Terrorism* 40, no. 3 (2017), 211–231. https://doi.org/10.1080/1057610X.2016.1188533.

[122] Haugstvedt, "A Flying Threat Coming to Sahel and East Africa? A Brief Review", 2020; Gettinger, D., *The Drone Data Book*, (Center for the Study of the Drone, Bard College, 2019), https://dronecenter.bard.edu/files/2019/10/CSD-Drone-Databook-Web.pdf.

[123] Schwartz et al., "Do Armed Drones Counter-Terrorism, or are they Counterproductive? Evidence from Eighteen Countries", 2022.

[124] Horowitz, et al., "Separating Fact from Fiction in the Debate over Drone Proliferation", 2016.

within the realm of outliers lies the inevitable spread of technology, with state actors continuously seeking to procure new advancements as they are developed. This trend parallels historical patterns observed with other military technologies, such as aircraft and tanks, which have undergone cycles of production, use, and eventual replacement.

Drones are crucial in facilitating real-time observation, power projection, intelligence-gathering, reconnaissance, surveillance and target cueing for battlefield missions. It provides both state and non-state armed groups with a potent and precise method for gathering intelligence and conducting lethal operations from a distance. Their versatility is evident as they are vital components in ISR missions.[125,126] Yet, the widespread proliferation of commercial and hobbyist drones in Africa carries unintended consequences that remain largely unexplored regarding their security implications. The recent announcement of an EU training mission to assist the Mozambican army in utilising drones to track militant activities underscores the accelerating growth of drone Africa, which encompasses both military and commercial unmanned aerial systems alongside consumer-grade drones.

Despite its negative security implications, the proliferation of drones also brings several advantages. Drones play a significant role in reshaping the security landscape by enabling the targeted elimination of high-value targets, facilitating power projection against terrorists and armed groups, providing crucial support to ground troops, and conducting vital intelligence, reconnaissance, and surveillance missions.[127] Moreover, they reduce the need for soldiers on the battlefield, particularly in coercive air campaigns against violent non-state actors operating in remote and inaccessible areas. Additionally, advancements in drone technology have spurred innovation in smart policing and expanded their applications beyond military contexts to include humanitarian efforts, such as coordinating law enforcement and domestic security services.[128]

However, the political economy of drone proliferation also presents several drawbacks. Firstly, it disproportionately empowers external forces by granting them access to and facilitating the pursuit of their strategic interests through drone engagement, whether through donations or support.[129] Moreover, when states procure drones from these suppliers, they become entangled in long-term dependencies for equipment replacement, spare parts, and upgrades.[130] This creates a bargaining chip for the supplying states to exert influence over the sale of their

---

[125] Boyle, M.J., *The Drone Age: How Drone Technology Will Change War and Peace* (Oxford: Oxford University Press, 2020).
[126] Gusterson, H., *Drone: Remote Control Warfare* (Cambridge: MIT Press, 2016).
[127] McCormick, "Exclusive: U.S. Operates Drones from Secret Bases in Somalia", 2 July 2015.
[128] Okpaleke, *Drones and US Grand Strategy in the Contemporary World*, 2023.
[129] Gatopoulos, A., "Largest Drone War in the World: How Airpower Saved Tripoli", *Al Jazeera*, 28 May 2020, www.aljazeera.com/news/2020/5/28/largest-drone-war-in-theworld-how-airpower-saved-tripoli.
[130] Turse, N., "Remote Warfare and Expendable People", *TomDispatch*, 30 January 2024, https://tomdispatch.com/remote-warfare-and-expendable-people/.

technology and perpetuates a cycle of dependency. This echoes historical drone colonial dynamics in the geopolitical demand and supply chain, further solidifying the selling country's leverage over its clients—African states.

Additionally, since these drones are often not locally manufactured, their servicing and maintenance typically require foreign expertise, perpetuating a cycle of dependence and highlighting Africa's peripheral status in a neo-colonial framework.[131] Moreover, the establishment of drone bases can reinforce the activities of authoritarian regimes in host countries, such as Idriss Déby in Chad and Ismaïl Omar Guelleh in Djibouti, who prioritise security cooperation and regime stability over addressing internal dissent and supporting rebel groups.[132]

Art[133] outlines various political purposes for using force, including compellence, projection, deterrence, and staggering. Compellence aims to enforce compliance through sheer force, while projection showcases force and strength. Deterrence seeks to prevent action or dissuade enemy targets from taking hostile actions due to the fear of military confrontation. Staggering, on the other hand, involves overwhelming enemy targets with brute force. These concepts are increasingly evident in the evolving landscape of dronification across Africa, which has assumed a neo-colonial character, where drones serve as tools for military sales and as instruments of statecraft and diplomacy.[134] The geopolitical implications of this phenomenon are both layered and multidimensional. The layered perspective acknowledges that geopolitics is inherent to international relations, encapsulating the strategic motives behind state actions to safeguard national interests and exert influence. This perspective emphasizes the importance of both hard and soft power capabilities in asserting influence on the global stage. Furthermore, the multidimensional nature of geopolitics recognizes that states rarely act alone but rather in collaboration with others within a dynamic ecosystem of power and influence.[135] While these interactions are not governed by a simple binary process, they reveal underlying contestations rooted in historical, military, strategic, and grand strategic factors that surpass superficial analysis. Moreover, the inevitability of technological spread drives state actors to continually seek and adopt new advancements, akin to historical cycles observed with other military technologies like aircraft and tanks. This evolution highlights the enduring quest for innovation in military arsenals, marked by a constant cycle of production, deployment, and replacement.

[131] Cannon, "Out of Africa: The Impact of Drones in Sub-Saharan Conflicts", 2023.

[132] Anyadike, "A Rough Guide to Foreign Military Bases in Africa", *The New Humanitarian*, 15 February 2017.

[133] Art, R.J. and Waltz, K.N. (Eds.), *The Use of Force: Military Power and International Politics* (Lanham: Rowman and Littlefield, 2009).

[134] Gettinger, "The Drone Data Book", 2019.

[135] Gonçalves, "The Scramble for Africa Reloaded? Portugal, European Colonial Claims and the Distribution of Colonies in the 1930s", 2021.

9 Conclusion

The proliferation of drones in Africa represents a multifaceted phenomenon with far-reaching implications for security, geopolitics, and state dynamics on the continent. As outlined in the preceding analysis, various state and non-state actors, including China, Iran, Turkey, and the United States, are actively involved in the expansion of drone technology across Africa. This expansion is driven by a complex interplay of economic, political, and strategic interests, with drones serving as instruments for achieving diverse objectives ranging from surveillance and reconnaissance to combat operations.

The implications of this drone proliferation are profound. On one hand, drones offer African states new capabilities for surveillance, counterterrorism, and border control, potentially enhancing their security and defense capabilities. However, the influx of drones also raises concerns about sovereignty and drone colonialism, as external actors gain greater access to African airspace and territories. Moreover, the reliance on foreign drone technology creates dependencies that may compromise African states' autonomy and decision-making processes.

Furthermore, the burgeoning proliferation of drones in Africa is presenting unmistakable signs of an imminent security dilemma, poised to revolutionize the region's security landscape by placing advanced technology within the reach of new actors, including violent non-state entities. However, the growing concern over the security vulnerabilities and threat susceptibilities posed by drones across Africa carries both metaphorical and paradoxical implications.

Metaphorically, the diffusion of drone technology has become pervasive, with an increasing number of states either acquiring or in the process of acquiring drones. This widespread adoption signifies the technology's transformative potential, ushering in a new era of military capabilities and strategic manoeuvrability across the continent.

Paradoxically, however, these efforts, ostensibly aimed at countering terrorism or enhancing military prowess, carry the inherent risk of unintended consequences, including heightened insecurity and civilian casualties. Furthermore, the spread of drones has implications for regional stability and conflict dynamics. Violent non-state actors, including terrorist groups like Boko Haram and al-Shabaab, are increasingly using drones for tactical purposes, further complicating security challenges in conflict-affected regions. Additionally, the emergence of drone-enabled proxy wars, exemplified by Iran's involvement in Sudan and Yemen, underscores the potential for drones to exacerbate existing conflicts and rivalries in Africa.

One prominent aspect of this trend is the emergence of drone diplomacy, where foreign actors leverage drone sales and technology transfer as a means to advance their geopolitical agendas and expand their influence in Africa. China, Iran, and Turkey have capitalized on gaps in arms control regimes to increase their presence on the continent, challenging the traditional dominance of Western powers such as the United States and France. This has led to a geopolitical contestation in which African states become arenas for strategic competition and influence projection among external actors.

In response to these developments, African states are faced with the challenge of navigating a rapidly evolving security landscape while balancing competing interests and priorities. There is a growing need for effective governance frameworks and regulatory mechanisms to mitigate the risks associated with drone proliferation, including the potential for misuse, proliferation of armed drones, and violation of human rights and international law. Moreover, African states must consider the broader geopolitical implications of drone proliferation and actively engage in regional and international dialogue to safeguard their interests and sovereignty. This includes strengthening diplomatic ties, enhancing cooperation on arms control, and leveraging partnerships to promote responsible and transparent use of drone technology.

While drones offer new opportunities for African states to address security challenges and enhance their capabilities, they also pose risks and challenges that must be addressed through concerted efforts at the national, regional, and international levels. Moving forward, African states must navigate this evolving landscape with vigilance and foresight to ensure that drones are harnessed as tools for peace, security, and development rather than instruments of conflict and domination and also take concerted measures to prevent the unchecked proliferation of drone bases across Africa. Instead, the focus needs to be directed to disarmament, demobilisation, and reintegration to shift efforts towards lasting solutions beyond military means particularly when neo-colonial motives are embedded.

Francis Okpaleke holds a Ph.D. in International Relations and Security Studies from the University of Waikato, New Zealand, an M.Sc. in Terrorism and Security Studies from the University of Salford, United Kingdom, and a B.Sc. (First Class Honours) in Political Science from the University of Nigeria. He serves as a senior policy analyst for the Canadian government and is a research fellow in New Zealand. Over the past decade, Okpaleke has gained diverse professional experience across four continents—Europe, Africa, North America, and Oceania—particularly in research and policy development. His research focuses on the intersection of emerging disruptive technologies, drones, counterterrorism, and critical security studies. Okpaleke has published extensively in peer-reviewed, high-impact journals, including *African Security Review*, the *Journal of Strategic Security*, *Sustainability*, and *Energy Research & Social Science*. He has also contributed several book chapters and policy insights. He is the author of Drones and US Grand Strategy in the Contemporary World (Springer, 2023) and the forthcoming Drones in African Battlespaces: Acquisition, Manufacture, and Deployment (Routledge, 2025).

Samuel Oyewole is a Postdoctoral Fellow at the University of Pretoria, South Africa, and a lecturer at Federal University Oye-Ekiti, Nigeria. His research focuses on African affairs, military strategy, and international relations. His work has been published in leading journals, including *Studies in Conflict & Terrorism*, *African Security Review*, and *Defence Studies*. He is the author of Utilitarianism in Outer Space: Space Policy, Socioeconomic Development, and Security Strategies in Nigeria and South Africa (Springer, 2024) and has co-edited five books, including Boko Haram's Campaign of Terror in Nigeria (Routledge, 2021) and Armed Banditry in Nigeria (Palgrave Macmillan, 2024).

Lethal Yet Limited: Evaluating Male Drones in Sub-Saharan Africa's Conflict Zones

Brendon J. Cannon and Ash Rossiter

Abstract

This article examines the efficacy of Medium Altitude Long Endurance (MALE) drones, specifically the Turkish-made Bayraktar TB2, in Sub-Saharan African conflicts. While MALE drones have demonstrated effectiveness in various global conflicts, their impact in Africa is constrained by vast distances, challenging environments, and complex internal and cross-border conflict dynamics. Case studies from the region that include Niger and Ethiopia reveal that these factors blunt the operational capabilities of drones like the TB2. Despite these limitations, the absence of modern integrated air defenses in the region allows MALE drones to remain lethal precision weapons, particularly in targeting supply lines and isolated groups. However, the success of these drones depends on competent operation, their employment in sufficient quantities, and adequate support infrastructure, which are often lacking in the region. We conclude that while MALE drones offer tactical advantages, they are unlikely to significantly alter the course of conflicts in Sub-Saharan Africa. In short, they are not the game-changing weapons that some analysts claim them to be.

Over the last three decades, there has been a steady diffusion of drones, including armed varieties, with a growing number of states and non-state actors acquiring them. The days when only the most advanced militaries operated armed drones are long gone. In sum, armed drones are now one of the most striking features of

B. J. Cannon (✉) · A. Rossiter
Khalifa University, Abu Dhabi, UAE
e-mail: brendon.cannon@gmail.com

A. Rossiter
e-mail: arossiter@ra.ac.ae

W. S. Coetzee and D. Putter (eds.), *Drones in the African Battlespaces*, Southern Space Studies, https://doi.org/10.1007/978-3-031-87147-4_4

modern warfare. They are ubiquitous above terrestrial and maritime battlespaces and can lay claim to having dramatically altered the course of conflicts.[1]

One key reason for the rapid proliferation of armed drones is the increasing number of countries manufacturing these uncrewed systems and their willingness to export them. China, Turkey, and Iran have developed lower-cost military drone programs and are willing to sell their products to most interested buyers.[2] Turkey in particular has become a major drone exporter especially to militaries incorporating such systems into their arsenals for the first time.[3] The Turkish-made Bayraktar TB2, a medium altitude long endurance (MALE) drone, has exemplified this shift. While it has been surpassed by next-generation drones and to some degree by the advent of cheaper and more numerous one-way attack drones at the time of writing in 2024, the TB2 clearly demonstrated the power of MALE drones when it successfully destroyed missile systems, main battle tanks, and contributed to the conflict outcomes in Syria, Libya and the Caucasus.[4]

MALE drones like the TB2 have captured significant attention due to their diverse capabilities and widespread use. Because MALE armed drones have been arguably the most high-profile new entrant in a range of modern conflicts in the past few years, they are the unit of analysis in this chapter. These drones do not possess the same capabilities as highly sophisticated (and expensive) high-altitude long-endurance (HALE) armed drones, like America's MQ-9 Reaper, which became synonymous with precision strike operations during the Global War on Terror (GWOT).[5] Although typically slower, less powerful and with shorter ranges, a new breed of tactical MALE armed drones, typified by the TB2, is nonetheless a powerful addition to many states' arsenals. The standard payload configuration of the aforementioned TB2 includes an electro-optical (EO) camera module, an infrared (IR) camera module, a laser designator, a laser range finder (LRF) and a laser pointer. It has four hardpoints with provisions to carry combinations of munitions, most usually the MAM-C and MAM-L laser-guided smart bombs, but also long-range anti-tank missiles and rockets. The aerial platforms are part of a larger system consisting of a ground control station, ground data terminal, remote display terminal, advanced base with generator and trailer modules.

[1] Rogers, J., "The Second Drone Age: defining war in the 2020s", *Defense & Security Analysis* 39, no. 2 (2023), 256–259.

[2] Ewers, E.C., Fish, L., Horowitz, M.C., Sander, A. and Scharre, P., *Drone Proliferation* (Center for New American Strategy, 2017). On Iran's drone program, see: Lob, E., "A New Arms Race Frontier: The Emerging Drone Competition Between Iran and Its Regional Rivals", in *The Arms Race in the Middle East: Contemporary Security Dynamics* (Cham: Springer International Publishing, 2023), 67–87.

[3] Cannon, B.J., "Turkey's military strategy in Africa", in *Turkey in Africa*, E.E., Tepeciklioğlu, and A.O. Tepeciklioğlu, (Eds.). (Oxon: Routledge, 2021), 127–143.

[4] Rossiter, A. and Cannon, B.J., "Turkey's rise as a drone power: trial by fire", *Defense & Security Analysis* 38, no. 2 (2022), 210–229.

[5] Cullen, T.M., "The MQ-9 Reaper remotely piloted aircraft: Humans and machines in action" (Doctoral dissertation, Massachusetts Institute of Technology, 2011).

These features make the TB2 a powerful capability in the hands of skilled operators. The fact that it is relatively cheap compared to its more sophisticated cousins on the international market makes it an affordable option for many militaries. Indeed, the TB2 has been likened to the Toyota Corolla of armed drones: they are relatively simple and inexpensive to operate and maintain, are reliable, and come at a good price point.[6] This combination has made the TB2 drone quite popular among African countries. Yet, the display effect of the TB2's prowess and record of providing military advantage to Turkey in Libya and Azerbaijan during the Nagorno-Karabakh conflict galvanized interest among African elites and militaries. The TB2's producer, Turkish company Baykar Makina, has now successfully sold these drones to African nations such as Niger, Togo, Djibouti, Libya, Ethiopia, Morocco, and Tunisia, with reports that Senegal, Kenya, Rwanda, Angola, and Nigeria are all eyeing their purchase or have them already.[7]

Rather than using the TB2 against external rivals, African nations see the usefulness of these drones primarily for internal security purposes. July 2022 saw a consignment of TB2s delivered to Togo, for example, to help the West African state stem the infiltration of jihadist fighters moving south from Burkina Faso. While in May of the same year, Niger acquired half a dozen of these drones for use against insurgent groups in the Sahel region south of the Sahara Desert, and around Lake Chad.[8]

Despite the ubiquity of drones, questions remain as to whether they can achieve similar results in all conflict scenarios and environments. Several scholars, for example, question the effectiveness of MALE drones because these systems are easy to detect and vulnerable to air defence capabilities.[9] In environments with modern networked integrated air defences, these systems, as Jacquelyn Schneider points out, "are easily detected by radars at long ranges, and often do not fly high enough to evade high-calibre anti-aircraft artillery or infrared-guided surface-to-air missiles."[10]

The limitations of MALE drones in environments with robust air defence systems and other countermeasures, such as electronic warfare systems, are well understood. This is not to imply that these deficiencies should be surprising; armed

[6] Stein, A., "From Ankara with implications: Turkish drones and alliance entrapment", *War on the Rocks*, 15 December 2021, https://warontherocks.com/2021/12/from-ankara-with-implications-turkish-drones-and-alliance-entrapment/.

[7] Campbell, M., "Drone Proliferation Dataset", *Center for New American Security*, 10 September 2024, www.cnas.org/publications/reports/drone-proliferation-dataset. See also Mitzer, S. and Oliemans, J., "An International Export Success: Global Demand For Bayraktar Drones Reaches All Time High", *Oryx*, 2 September 2022, https://www.oryxspioenkop.com/2021/09/an-international-export-success-global.html.

[8] Melly, P., "Turkey's Bayraktar TB2 drone: why African states are buying them", *BBC*, 22 August 2022, www.bbc.com/news/world-africa-62485325.

[9] Calcara, A., Gilli, A., Gilli, M. and Zaccagnini, I., "Will the Drone Always Get Through? Offensive Myths and Defensive Realities", *Security Studies* 31, no. 5 (2022), 791–825.

[10] Schneider, J., "Unscorable at 12: Technically correct, but misses the mark", *Security Studies* 32, no. 3 (2023), 569.

MALE drones were not designed to be used in settings where control of the air was heavily contested. They came to prominence in the GWOT in Pakistan and Afghanistan, where the U.S. could assume air supremacy against insurgents.

Beyond contested airspace, it is important to emphasise—though it is often underappreciated—that MALE drones display notable limitations when deployed across extreme or challenging terrain. This issue is compounded when these systems are operated by states and armed forces lacking the capabilities to fully exploit the drones to their full advantage. In the worst cases, inexperienced operators can make tactical errors that result in civilian casualties.

To explore this argument, the chapter examines the limitations of MALE drone use in conflicts where the targets lack modern air defence systems. By removing air defences as a primary constraint, we highlight other factors that limit the effectiveness of these systems. To do so, we focus on the efficacy of MALE drones in recent Sub-Saharan African conflicts, where vast distances, harsh environmental conditions, and complex internal and cross-border dynamics presented significant challenges to operators. The case studies demonstrate that Sub-Saharan Africa's operational environment—marked by difficult terrain and logistical hurdles—frequently blunted the effectiveness of MALE drones and limited their operational reach. These findings, however, have broad applicability to other extreme environments around the world.

The findings also reveal that despite some limitations, MALE drones—when operated effectively and in sufficient numbers—were formidable precision weapons in African skies. The lack of modern air defence systems among adversaries in the region allowed these drones to operate with near impunity and made them valuable assets for states and political/military leaders facing insurgencies and asymmetric warfare. Yet, the true potential of MALE drones in Sub-Saharan Africa depended heavily—and will continue to depend—on overcoming logistical challenges, deploying adequate support infrastructure, and ensuring competent operation. These will remain critical for maximizing drone effectiveness in this unique and demanding environment for the foreseeable future.

1 Terrain, Distance and Operators

1.1 Terrain

Topographical and environmental conditions across Sub-Saharan Africa continue to complicate the deployment and utility of armed drones. The landscape varies dramatically, from the low-lying Lake Chad Basin in the Sahel to the towering Ethiopian highlands, where altitudes often exceed 1500 m and can reach as high as 4550 m. These high altitudes strained the performance of drones like the TB2, particularly when operating close to their maximum altitude limits. Additionally, weather can be a significant, impeding factor in terms of MALE drone performance. Frequent dust and sandstorms originating from the Sahara Desert can severely impair the drones' visible-light sensors and reduce their effectiveness. Similarly, dense cloud cover is common in regions like highland Ethiopia, the Gulf of Guinea littorals, and Central Africa. The latter region, with some of the world's densest forest canopies, provides ample cover for insurgents, further challenging drone operations.

1.2 Distance

Sub-Saharan Africa is a vast and diverse region, covering over 21 million square kilometers, or about 20% of the earth's surface. To put this in perspective, the subcontinent of India, if superimposed, would barely cover the Horn of Africa.[11] Given such immense distances, the operational range of armed drones like the Turkish-made Bayraktar TB2 continues to present significant challenges. Although the TB2 is equipped with a maximum range of around 300 km and a loitering time of 27 h when lightly armed, these capabilities are, at times, insufficient in expansive terrains like Ethiopia or Niger in the Sahel region. For example, TB2 drones delivered by Turkey to Ethiopia to combat Tigrayan militias were reportedly based in locations such as Harar Meda[12] near the capital, Addis Ababa. However, the distance from these bases to Tigray's capital, Mekele, exceeded the TB2's effective range. The Ethiopian government had to reportedly move the TB2s north to the Bahir Dar airport in the Amhara Region.[13] Only from this civilian airport, which doubles as a military airbase, could the TB2s with their 300-km range reach many of the Tigray People's Liberation Front (TPLF) positions. It was from here that a TB2 was likely launched that on 7 January 2022 struck a camp for internally displaced people in the town of Dedebit in the far north of the country near the border of Eritrea. The Ethiopian government denied it was responsible for the attack which reportedly killed nearly 60 civilians.[14] A drone that matched the measurements and description of the Turkish TB2, however, was spotted on the tarmac of Bahir Dar Airport the day before the strike. In addition, Ethiopia was the only party in the conflict operating armed drones at this point.

Despite these challenges, MALE drones like the TB2 retain some advantages in these environments due to their electro-optical and infrared (EO/IR) payloads,

[11] Visual Capitalist, "Map of the True Size of Africa", www.visualcapitalist.com/map-true-size-of-africa/.

[12] Google Earth imagery (February 2022; 8° 42′ 31.48″ N; 39° 0′ 18.00″ E) showed two possibly Chinese-made Wing Loong MALE drones at Harar Meda and a report placed a Wing Loong I MALE drone there on 3 December 2022. Twitter, https://twitter.com/wammezz/status/1600453844735234049, last accessed 2024/09/04.

[13] Google Earth imagery (10 June 2022) showed one TB2 at the Bahir Dar airfield.

[14] Bearak, M., Kelly, M. and Lee, J.S., "How Ethiopia used a Turkish drone in a strike that killed nearly 60 civilians", *Washington Post*, 7 February 2022, www.washingtonpost.com/world/interactive/2022/ethiopia-tigray-dedebit-drone-strike/.

which offer a 360° view and allow the drones to operate in diverse weather conditions. They can fly at altitudes as high as 4500 m, or "under the weather," as needed. However, operating at lower altitudes to avoid weather conditions or enhance target visibility brings its own risks, as it exposes the drones to potential small-arms fire. Moreover, as discussed above, Sub-Saharan Africa's huge distances limited the optimal usage of these drones, as the range and loitering capabilities were often insufficient to cover the expansive operational areas required. This remains the case today with MALE drones, although newer drones have longer range and increased payloads. Burkina Faso and Ethiopia have followed up on TB2 purchases with the Bayraktar Akıncı HALE armed drone, also made by Turkish company Baykar.

To mitigate these limitations, various solutions were operationalized, such as the use of relay masts that could be towed and deployed to extend the range of the TB2.[15] Establishing airstrips closer to the conflict zones was another potential solution that could have enabled the drones to reach insurgents operating in remote areas. Additionally, drone systems like the TB2 can be containerized and transported to forward positions.[16] However, in regions like the Sahel, Somalia, and South Sudan, where roads are often non-existent or in poor condition, particularly in remote and difficult terrain, containerization and road transport become costly and challenging. The necessity of a forward airfield further complicates the logistics, making the forward deployment and sustained operation of armed drones in such environments potentially too costly and inefficient, thereby reducing their overall effectiveness.

1.3 Operators

The effectiveness of armed drones is not only a matter of technology and logistics but also heavily dependent on the competence of their operators. Purchasing a drone system is one thing, but effectively operating it is another. While the usage of one-way attack drones in Ukraine has shown that mass coupled with some degree of accuracy is useful for destructive attacks against infrastructure, the recent use of MALE drones in Ethiopia, Ukraine, and Yemen, among other conflicts, has shown that poor operation can render even the most advanced drones ineffective. For example, Turkey reportedly lost ten TB2 drones in Libya within a few months due to operational errors, prompting the deployment of experienced operators from the

[15] Düz, S., "The Ascension of Turkey as a Drone Power: History, Strategy, and Geopolitical Implications", *SETA Analysis 65* (Ankara: Foundation for Political, Economic and Social Research, 2020).

[16] Each Bayraktar TB2 system consists of six aerial vehicles, two ground control stations (GCS), three ground data terminals (GDTs), two remote video terminals (RVTs) and ground support equipment. The UAV has a length of 6.5 m, wingspan of 12 m and a maximum take-off weight of 650 kg. (Army Technology, "Bayraktar TB2 Tactical UAV", *Army Technology*, 6 January 2021, www.army-technology.com/projects/bayraktar-tb2-tactical-uav/).

manufacturer, Baykar Makina.[17] These operators were able to use the drones more effectively, penetrating Libya's relatively lightly defended airspaces and successfully targeting Libyan National Army (LNA) surface-to-air launchers and radars.[18] It is likely that Turkish operators also supported Azerbaijani forces in their conflict, demonstrating the importance of skilled operation.[19] However, such support is driven by Turkey's national interests, and it is unlikely that Ankara would extend this level of involvement to conflicts in Sub-Saharan Africa, where the stakes are not as high for Turkish national security.[20]

Poor operator abilities and failure to properly maintain the systems will likely degrade Sub-Saharan drone fleets over the medium term. In 2023, for example, a Bayraktar TB2 drone belonging to the Burkinabe Air Force crashed approximately 140 km southeast of the capital city, near Tenkodogo.[21] Technical failures and accidents can occur with any aircraft, manned or unmanned. Nonetheless, the crash is undoubtedly a setback for Burkina Faso with its small armed forces facing multiple insurgencies.

To add another layer of complexity, drone systems should be understood as part of a larger reconnaissance strike complex. A reconnaissance strike complex is made up of several key technologies rather than a single technology. Modern militaries possess intelligence, surveillance, and reconnaissance (ISR) assets that locate and track potential targets and then send strike platforms that deliver munitions against these targets. Whilst drones such as the TB2 are ISR assets with onboard munitions, they are most efficiently used in intelligence-led operations that leverage a state's full array of capabilities to locate and fix targets. This point is especially apposite in the context of Sub-Saharan conflicts, given the vast distances involved and dispersed nature of the anti-government forces.

When the proliferation of armed drones combines with structural flaws in intelligence gathering and poor targeting processes, tragedy can be the result. In December 2023, a Nigerian military drone strike accidentally struck a village in northwest Kaduna State, killing 85 civilians as they celebrated a Muslim festival.[22]

[17] Calcara, A., Gilli, A., Gilli, M., Marchetti, R. and Zaccagnini, I., "Why drones have not revolutionized war: The enduring hider-finder competition in air warfare", *International Security* 46, no. 4 (2022), 130–171.

[18] Rossiter, A. and Cannon, B.J., "Game-Changing Drones? The Record from Libya to Ukraine", in *De Gruyter Handbook of Drone Warfare*, ed. J. Rogers (2024).

[19] Kasapoğlu, C., "Turkey Transfers Drone Warfare Capacity to its Ally Azerbaijan", *Eurasia Daily Monitor* 17, no. 144 (2022), https://jamestown.org/program/turkey-transfers-drone-warfare-capacity-to-its-ally-azerbaijan/.

[20] Cannon, B.J., "Turkey's military strategy in Africa", in *Turkey in Africa* (Oxon: Routledge, 2021), 127–143.

[21] Lionel, E., "Burkinabe Air Force Bayraktar TB2 drone crashes near Tenkodogo", *Military Africa*, 22 June 2023, www.military.africa/2023/06/burkinabe-air-force-bayraktar-tb2-drone-crashes-near-tenkodogo/.

[22] Assherton, R., "Nigerian army drone strike accident kills at least 85 civilians", *Guardian*, 5 December 2023, www.theguardian.com/world/2023/dec/05/nigeria-army-drone-strike-accident-kills-civilians.

A Nigerian army unit engaged in a firefight had reportedly given the wrong grid reference to the drone operator.

In conclusion, while armed drones like the TB2 offer significant tactical advantages in specific scenarios, their effectiveness in Sub-Saharan Africa is constrained by a combination of vast distances, challenging environments, and the need for competent operators and effective employment. These factors can blunt the power of drones, limiting their operational reach and impact. Nonetheless, when used strategically and in sufficient numbers, drones remain a potent tool, capable of providing precision strikes and valuable intelligence in conflicts where adversaries lack sophisticated air defences.

2 Going on the Offensive

Despite the challenging environmental factors in Sub-Saharan Africa, MALE drones can still perform effectively in offensive operations, largely due to the region's limited air defence capabilities. To date, few African states possess sophisticated air defence systems. This is changing, however, as states like Kenya and Rwanda are reportedly looking to purchase such systems. In June 2024, the Kenya Defence Forces (KDF) was exploring the purchase of Israel's $4 billion Raphael Spyder advanced air defence system, in part to guard against unmanned aerial attacks.[23] In early 2024, Russia deployed and installed air defence systems in Niger to assist its military-led government that had removed President Mohamed Bazoum and his democratically elected government in 2023, according to media reporting.[24] South Africa, which has traditionally had the most advanced defence industry on the continent, has developed the Umkhonto ground-based launcher (GBL), a mobile surface-to-air missile (SAM) vertical launching system. The missile system is intended to provide all-round protection for the armed forces against airborne threats, including attack aircraft, missiles and unmanned aerial vehicles (UAVs). These examples are exceptions. As far back as 2019, Rwanda Defence Force Air Force Chief of Staff, Maj Gen Charles Karamba, highlighted the paucity of air defence systems on the continent., "Many countries have made modernisation of their military technologies as one of the four pillars of their military capability alongside force structure, readiness and sustainability. Air defence, a

[23] Mutua, J., "Kenya's targeted Israeli air defence system to cost at least Sh4 billion", *Business Daily*, 17 June 2024, www.businessdailyafrica.com/bd/economy/kenya-s-targeted-israeli-air-defence-system-to-cost-sh4-billion--4660476.

[24] Al Jazeera, "Russia sends military trainers, air defence system to Niger: State media", *Al Jazeera*, 12 April 2024, www.aljazeera.com/news/2024/4/12/russia-sends-military-trainers-air-defence-system-to-niger-state-media.

branch of military modernisation is seen as bridge to success in controlling diverse threats for common interest," he said.[25]

The need for advanced air defence systems is certainly growing as the availability of drones and especially one-way attack drones increases in line with their costs decrease and accessibility.[26] Yet, for states that have acquired MALE drones in Sub-Saharan Africa, they remain formidable not so much as weapons (as will be seen below), but as ISR tools. This is because traditionally, non-state armed groups, such as al-Shabaab in Somalia, ISIS in Chad, and the Tigrayan forces in Ethiopia, have generally lacked the necessary equipment, such as man-portable air-defence systems (MANPADS) or advanced anti-aircraft guns to effectively counter such drones. Even if these groups did acquire them, MALE drones like the TB2 can operate above their effective range, allowing them to loiter at high altitudes while surveilling and targeting their adversaries. This gives countries like Niger, which recently acquired TB2 drones, a significant tactical advantage over lightly armed insurgents.

In April 2023, for example, Ankara reportedly sold Bayraktar TB2 drones to Somalia as part of counterterrorism efforts. The Somali Federal Government in Mogadishu had no doubt hoped that these drones would provide them with a tactical edge over al-Shabaab, which had erased many of the gains made by Somali National Army and African Union troops, as well as numerous local militias.[27] Yet, even supported by TB2s and U.S. military drones, al-Shabaab continues to hold and even gain territory and foment terrorist attacks in Mogadishu. More worrying, the terrorist group that already has ties to al-Qaeda and ISIS is assessed to be now working with Houthi militants to acquire more weapons and expand its capabilities.[28]

The example above raises another important point. The rapid proliferation of MALE drones in Sub-Saharan Africa from about 2021 onwards provided governments, for a time, with a new advantage over terrorists, insurgents, separatists and cross-border armed groups. Yet we know that when a new weapon is introduced into conflict, targets will try to adapt by adopting the same weapon or, if they cannot do so, develop ways to blunt or repel the weapon. As such, it can be expected that non-state armed groups, whether in Mali, Somalia, or Nigeria, will find their

[25] Rwanda MOD, "Maj Gen Karamba highlights the new era for African air defences at Aviation Africa Summit", Rwandan Ministry of Defence, 27 February 2019, www.mod.gov.rw/news-detail/maj-gen-karamba-highlights-the-new-era-for-african-air-defences-at-aviation-africa-summit.

[26] Plichta, M. and Rossiter, A., "A one-way attack drone revolution? Affordable mass precision in modern conflict", *Journal of Strategic Studies* 47, no. 6–7 (2024), 1001–1031.

[27] Baez, K., "Turkey signed two major deals with Somalia. Will it be able to implement them?", *Atlantic Council*, 18 June 2024, www.atlanticcouncil.org/blogs/turkeysource/turkey-signed-two-major-deals-with-somalia-will-it-be-able-to-implement-them/.

[28] Hochet-Bodin, N., "Somalia: Breakthrough in the war against Al-Shabab", *Le Monde*, 20 November 2022, www.lemonde.fr/en/le-monde-africa/article/2022/11/20/somalia-breakthrough-in-the-war-against-al-shabab_6004993_124.html.

own ways to adopt an emerging technology. Al-Shabaab, for example, has repeatedly used drones to conduct surveillance on African Union troops and the Somali National Army's bases and convoys.[29] On the other side of the continent, in July 2022, in the town of Gubio, Nigeria, the Islamic State West African Province (ISWAP) used a surveillance drone to survey the location of a Nigerian military convoy before ambushing it. It might not be long before African non-state armed actors start using explosive-laden drones in kamikaze attacks[30] or adapt commercial off-the-shelf platforms to carry and drop small military-grade munitions—as the Islamic State did in Iraq during the battle for Mosul.[31] African militant groups like Boko Haram, ISWAP and al-Shabab are acquiring and modifying drones for their own purposes. Other non-state armed actors may not be far behind. In the future, drone proliferation—especially the spread of military-grade drones to non-state actors—could change the balance of power between governments and their opponents. Hence the importance of arms control.

For now, the drone wars in Sub-Saharan Africa remain highly asymmetric, favouring those governments that have acquired armed MALE drones. The TB2's ability to carry up to four precision-guided MAM-L or MAM-C munitions gives possessors of this system quite formidable offensive capabilities relative to their opponents.[32] These munitions are ideal for targeting small vehicles, unhardened structures, or groups of soldiers, making the TB2 highly effective for decapitation strikes, neutralising key combatants, or destroying critical materiel such as main battle tanks, armoured vehicles, and artillery. However, MALE drones are less suited for direct engagement with massed troops in large-scale conflicts, as seen in Ethiopia's civil war. In such scenarios, the Ethiopian government likely relied more on heavy payload-carrying fixed-wing attack aircraft like the SU-25s or MiG-23s to repel advances by the Tigray Defence Forces.[33] In contrast, MALE drones excel in disrupting supply lines and targeting isolated forces, as demonstrated in Libya, where TB2 drones successfully attacked supply columns heading toward Tripoli, even neutralising Russian-made Pantsir-S1 anti-aircraft systems in the process.[34]

Given that most conflicts in Sub-Saharan Africa have not involved large-scale troop formations or clearly defined battlefronts, MALE drones continue to offer a

[29] Haugstvedt, H., "A flying threat coming to Sahel and East Africa? A brief review", *Journal of Strategic Security* 14, no. 1 (2020), 92–105.

[30] Plichta and Rossiter, "A one-way attack drone revolution? Affordable mass precision in modern conflict", 2024.

[31] Rossiter, A., "Drone usage by militant groups: exploring variation in adoption", *Defense & Security Analysis* 34, no. 2 (2018), 113–126.

[32] Carrying a full payload lessens flight time to 14 h. (Times Aerospace, "Libya's deadly game of drones", n.d., https://www.timesaerospace.aero/features/defence/libyas-deadly-game-of-drones).

[33] Cannon, B.J., "Out of Africa: the impact of drones in Sub-Saharan conflicts", *Defense & Security Analysis* 39, no. 2 (2023), 267–270.

[34] Rossiter, A. and Cannon, B.J., "Turkey's rise as a drone power: trial by fire", *Defense & Security Analysis* 38, no. 2 (2022), 210–229.

good return on investment for African states. These drones are particularly well-suited for the type of low-intensity, irregular- and asymmetric warfare common in the wider region, where they can deliver precise strikes against dispersed enemy forces without facing significant threats from air defences. Ethiopia's Tigray War, which involved more traditional military engagements, is somewhat of an anomaly in this context, resembling interstate conflicts like the Ogaden War of 1977–78. For most African states, the acquisition of armed drones represents a strategic asset that can provide a critical edge in ongoing and future conflicts.

3 Conclusion

The effectiveness of MALE armed drones in Sub-Saharan African combat scenarios is significantly influenced by the vast distances and challenging environmental conditions that characterise the majority of the continent. While drones can deliver substantial tactical benefits by targeting enemy supply lines, artillery, and small groups of fighters, they face limitations in dealing with massed infantry and in holding territory—challenges that are not unique to drones but apply to air power in general. Additionally, the fluid, hit-and-run tactics and adaptable strategies employed by many non-state armed groups in Africa further complicate the drones' ability to decisively alter the outcome of conflicts.[35]

Armed drones, such as the TB2, equipped with advanced EO/IR sensors, laser targeting, and precision munitions, can still be highly effective within a certain operational range, particularly in the low-intensity, asymmetric conflicts that are common across the continent. These drones provide a persistent aerial presence, offering Sub-Saharan African military leaders a powerful tool to monitor and strike at enemy forces from a distance. However, this advantage is tempered by the fact that drones alone are unlikely to dramatically change the course of these conflicts. The vast and varied geography, combined with the decentralised and mobile nature of many insurgent groups, presents a significant challenge that drones cannot easily overcome.

The strategic value of drones in Sub-Saharan Africa also depends heavily on the availability of skilled operators and the necessary support infrastructure. While drones like the TB2 can offer significant advantages, their effectiveness is limited without enough trained personnel and logistical support to maintain and deploy them effectively across the vast African landscape. Until African states can invest in and develop these capabilities at scale, the impact of drones on the region's conflicts will remain limited.

Cost is another critical factor that affects the proliferation and use of drones in Sub-Saharan Africa. Although a $5 million drone like the TB2 might seem like a cost-effective solution, this perception is relative to the financial resources

[35] On the debate over drones' game-changing effects in conflict, refer to Rossiter, A., "Military technology and revolutions in warfare: priming the drone debate", *Defense & Security Analysis*, 39, no. 2 (2023), 253–255.

available to each state. For instance, Ethiopia, which reportedly had only 20 days of currency reserves in late 2022, may find it challenging to sustain an effective drone program. The financial constraints faced by many African nations mean that, despite the growing number of drones in their arsenals, insurgents may still retain a tactical advantage in many areas. The high cost of acquisition, operation, and maintenance of drones poses a significant barrier to their widespread adoption and effectiveness in the region's complex and resource-strained conflict environments.

Acknowledgements The author would like to thank Olli Pekka Suorsa for his helpful comments and insights.

Brendon J. Cannon is an Associate Professor of International Security at Khalifa University of Science and Technology in Abu Dhabi, UAE. His research lies at the intersection of security studies, classical geopolitics, and transformative technologies. He has published on various topics related to defence industries, the effectiveness of armed drones as strategic and tactical weapons, and the shifting distributions of power across the Indo-Pacific. Cannon's articles have appeared in *Defence Studies, Small Wars and Insurgencies, International Politics,* and *Third World Quarterly.* His book, edited with Kei Hakata, is titled *Indo-Pacific Strategies: Navigating Geopolitics at the Dawn of a New Age* (Routledge). He earned his Ph.D. in Political Science from the University of Utah, USA, in 2009.

Ash Rossiter serves as the Lead Researcher and Associate Professor specialising in Defence and Security at Rabdan Academy in Abu Dhabi. He earned his Ph.D. from the University of Exeter in 2014 after obtaining an MA in War Studies from King's College London. Rossiter has extensively published on topics related to technology and international security, strategy, warfare, and comparative defence industries. His research appears in peer-reviewed journals, including the *Journal of Strategic Studies*, *International Relations*, *Intelligence & National Security*, *Defence Studies*, and *Parameters*. He co-authored the book Warfare in the Robotics Age with Peter Layton, published in 2024 by Lynne Rienner.

France's Involvement in Drone Warfare in Mali

Arinze Ngwube

Abstract

France's involvement in drone warfare in Mali forms a crucial aspect of its broader counterterrorism strategy in the Sahel region, particularly through Operation Barkhane. This military operation, initiated in 2014, aims to combat insurgent groups such as Al-Qaeda in the Islamic Maghreb (AQIM) and the Islamic State in the Greater Sahara (ISGS). Drones have been pivotal to France's military efforts, providing real-time intelligence, surveillance, reconnaissance (ISR), and precision strikes. The integration of drones, primarily MALE (Medium Altitude Long Endurance) systems like the MQ-9 Reaper, has enabled France to conduct targeted operations against terrorist leaders and disrupt militant networks while minimising the risk to ground troops. France's use of drones has proven vital for monitoring vast, remote areas and responding rapidly to emerging threats. However, this reliance on drone warfare has raised concerns over civilian casualties, legal implications, and the long-term effectiveness of counterinsurgency tactics. Furthermore, the growing reliance on UAVs highlights France's broader shift towards modernised, technology-driven warfare in its former colonies. As the security situation in Mali and the wider Sahel remains precarious, the role of drones in France's military strategy continues to evolve, reflecting both operational successes and the challenges of asymmetrical warfare.

A. Ngwube (✉)
Department of Political Science, Federal University Oye-Ekiti, Oye, Ekiti, Nigeria
e-mail: ngwube@hotmail.co.uk

W. S. Coetzee and D. Putter (eds.), *Drones in the African Battlespaces*, Southern Space Studies, https://doi.org/10.1007/978-3-031-87147-4_5

1 Introduction

Since 2012, Mali has been engulfed in a conflict that has jeopardised not only its stability but also the broader security of the Sahel region. The crisis began with a Tuareg rebellion in northern Mali, which rapidly escalated as Islamist militant groups seized control of vast areas. The deteriorating security situation quickly became an international concern, prompting France to intervene militarily in 2013 with Operation Serval, aimed at halting the advance of militants and restoring Mali's territorial integrity. This mission evolved into Operation Barkhane in 2014, expanding France's counterterrorism efforts across the Sahel.

France's involvement in Mali's conflict, particularly through the use of drones, represents a significant evolution in its military engagement strategy. Drone warfare has emerged as a pivotal component of modern military operations, offering substantial advantages in surveillance, intelligence gathering, and precision strikes. The use of Unmanned Aerial Vehicles (UAVs) has allowed France to monitor vast, remote areas and respond swiftly to emerging threats, minimizing the risk to ground troops. The introduction of drones, particularly MALE (Medium Altitude Long Endurance) systems such as the MQ-9 Reaper, has significantly enhanced France's operational capabilities in Mali. These drones provide persistent monitoring and the ability to conduct targeted operations against terrorist leaders, disrupting militant networks with precision strikes. However, the increasing reliance on drone warfare raises significant ethical, legal, and strategic concerns. Issues surrounding civilian casualties, sovereignty violations, and the long-term effectiveness of counterinsurgency strategies in the region persist.

This chapter aims to explore France's use of drones in Mali by examining its strategic objectives, the impact on counterterrorism efforts, and the broader implications for regional security and international norms. Through this analysis, we can gain a deeper understanding of the complexities of modern warfare in an asymmetrical conflict zone like the Sahel, where France's technological advantages must be weighed against the ethical challenges and potential long-term effects on regional stability.

2 Historical Context of French Military Involvement in Mali

Mali, then known as French Sudan, was part of French West Africa, a federation of eight territories controlled by France from the late nineteenth century until its independence in the 1960s.[1] During the colonial period, France established administrative, economic, and military control over the region, leaving a legacy of centralized governance and Francophone culture that persists in Mali today. After

[1] Lansdorf. T., "Mali", in *Politcal Handbook of the world*, ed. T. Lansdorf (903–910). (Thousand Oaks, CA: CQ Press, 2012).

gaining independence in 1960, Mali maintained close ties with France, particularly in areas of political cooperation, economic assistance, and security. France continued to exert considerable influence in its former colonies, often acting as a guarantor of stability. This influence was formalized through defence agreements, economic partnerships, and diplomatic relations that ensured France's continued presence and strategic interests in the region.[2] Political instability, economic challenges, and internal conflict marked Mali's post-independence period. However, the situation took a more severe turn in the early 2000s, culminating in the 2012 crisis.[3] The fall of Libyan leader Muammar Gaddafi in 2011 had a profound impact on the Sahel region. Weapons from Gaddafi's arsenals and battle-hardened fighters, many of whom were Tuareg mercenaries, flowed into northern Mali.[4] This influx of arms and combatants fuelled a new Tuareg rebellion led by the National Movement for the Liberation of Azawad (MNLA), which sought independence for northern Mali.

The situation quickly escalated as Islamist militant groups, including Ansar Dine, Al-Qaeda in the Islamic Maghreb (AQIM), and the Movement for Oneness and Jihad in West Africa (MOJWA), took advantage of the chaos. These groups overpowered the MNLA and imposed strict Sharia law across the territories they controlled.[5] The Malian government, weakened by internal divisions and a military coup in March 2012, was unable to respond effectively to the crisis. Faced with the prospect of a failed state and the potential for a terrorist stronghold in the heart of West Africa, France launched Operation Serval in January 2013.[6] The operation was a rapid military intervention aimed at stopping the advance of Islamist militants towards the Malian capital, Bamako, and restoring the sovereignty of the Malian state. French forces, supported by Malian troops and later by African and international partners, succeeded in recapturing key northern cities such as Gao, Timbuktu, and Kidal.[7]

Operation Serval achieved its immediate objectives, but it also highlighted the complex and enduring nature of the conflict in Mali. Recognising the need for

---

[2] Kehinde, T., "French Colonial Policies in West Africa: Power Dynamics, Cultural Impositions and Economic Legacies", *International Journal of Advances in Social Sciences and Humanities* 3, no. 1 (2024), 8–17.

[3] Bruce, W., "Mali: Collapse and Instability", In *The Oxford Handbook of the African Sahel*, ed. L.A. Villalón (Oxford: Oxford University Press, 2021).

[4] Baudais, V., "The Impact of the Malian Crisis on the Group of Five Sahel Countries: Balancing Security and Development Priorities", *Stockholm International Peace Research Institute*, 18 May 2020, www.sipri.org/commentary/topical-backgrounder/2020/impact-malian-crisis-group-five-sahel-countries-balancing-security-and-development-priorities.

[5] Al Jazeera, "International Condemnation for Mali Coup", *Al Jazeera*, 23 March 2012, www.aljazeera.com/news/2012/3/23/international-condemnation-for-mali-coup.

[6] Boeke, S. and B. Schuurman, "Operation 'Serval': A Strategic Analysis of the French Intervention in Mali, 2013–2014", *Journal of Strategic Studies* 38, no. 6 (2015), 801–825.

[7] Flood, H. D., "A Review of the French Led Military Campaign in Northern Mali", *Combating Terrorism Center at West Point*, 2013, https://ctc.usma.edu/a-review-of-the-french-led-military-campaign-in-northern-mali/.

a sustained counterterrorism effort, France transitioned from Operation Serval to Operation Barkhane in August 2014.[8] Operation Barkhane expanded France's military presence across five Sahel countries—Mali, Niger, Chad, Burkina Faso, and Mauritania—reflecting a broader regional approach to combating terrorism and stabilising the region.[9] France's involvement in Mali, therefore, is not just a response to the immediate threat posed by Islamist militants but also a continuation of its long-standing role as a critical security actor in West Africa. The evolution of the conflict and the emergence of new security challenges, including the spread of extremism and the fragmentation of militant groups, have shaped France's military strategy, leading to the adoption of new tools and tactics, including drone warfare, to address these ongoing threats.

3 France's Adoption of Drone Warfare in Mali

Using drones in Mali has become a notable aspect of military operations, reflecting the integration of uncrewed aerial vehicles (UAVs) into modern conflict scenarios. These drones are employed for surveillance, reconnaissance, and targeted strikes, providing a strategic advantage in monitoring vast and often challenging terrains.[10] They are crucial in Mali's efforts to enhance security and counterterrorism and collect intelligence in remote or inaccessible areas. Using drones in Mali reflects a multifaceted approach, combining military and civilian applications to address security challenges. Introducing drones in Mali's military security landscape has sparked debates on their effectiveness, ethical considerations, and the potential to reshape the dynamics of asymmetric warfare.[11] This overview explores the various roles and implications of drone usage in Mali, shedding light on the opportunities and challenges of this evolving technology.

In response to the evolving security threats in Mali and the broader Sahel region, France has increasingly relied on drone warfare as a critical component of its military strategy.[12] The complex and vast terrain of the Sahel, coupled with the mobile

[8] Boeke, S. and A. Tisseran, "Mali's Long Road Ahead", *The RUSI Journal* 159, no. 5 (October–November 2014), 32–40.

[9] Ministère de la Défense, "French White Paper: Defence and National Security", 2013, www.defense.gouv.fr/english/content/download/206186/2393586/file/White%20paper%20on%20defense%20%202013.pdf.

[10] Candelmo, C., "Drones at War: The Military Use of Unmanned Aerial Vehicles and International Law", In *Use and Misuse of New Technologies*, ed. E. Carpanelli and N. Lazzerini (Cham: Springer, 2019), 93–112.

[11] Akande, D. and A. Leveringhaus, "Drone Strikes: Ethics and Laws of 21st Century Warfare", Oxford Martin School, University of Oxford, 27 February 2014, www.oxfordmartin.ox.ac.uk/events/drone-strikes-ethics-and-laws-of-21st-century-warfare/; Annis, F., "Drones and the Legality and Ethics of War", *Small Wars Journal*, 2 February 2020, https://smallwarsjournal.com/jrnl/art/drones-and-legality-and-ethics-war.

[12] BBC, "Why France is Focused on Fighting Jihadists in Mali", *BBC*, 27 November 2019, www.bbc.com/news/world-africa-50558972.

and elusive nature of extremist groups, made traditional military operations challenging.[13] To address these difficulties, France integrated UAVs into its operations, primarily through the use of MALE drones, such as the MQ-9 Reaper.[14] These drones, acquired from the United States, are equipped with advanced surveillance systems and precision-guided munitions, enabling France to conduct intelligence-gathering and targeted strikes with minimal risk to its personnel. The Sahel's vast, sparsely populated regions and the militants' ability to blend into the local environment required robust intelligence capabilities.[15] Drones provided French forces with real-time, high-resolution imagery and data, enabling them to track militant movements, identify potential threats, and plan operations with greater precision. This capability was crucial in a region where ground intelligence was often limited or unreliable. The introduction of armed drones allowed France to carry out targeted strikes on high-value militant leaders and strategic assets. By focusing on key figures and infrastructure, France aimed to disrupt the command structures and operational capabilities of extremist groups like Al-Qaeda in the Islamic Maghreb (AQIM) and the Islamic State in the Greater Sahara (ISGS). The precision of drone strikes also helped minimise collateral damage, a critical concern given the proximity of militants to civilian populations.[16]

Drones also played a vital role in protecting French and allied forces during ground operations. They provided real-time reconnaissance, detected potential ambushes, and delivered air support when troops encountered resistance. This capability allowed French forces to operate more effectively and with reduced risk, particularly in remote or hostile environments where traditional air support might not be immediately available. France's drone operations in Mali were not standalone efforts but were integrated into a comprehensive military strategy under Operation Barkhane. France's adoption of Drones was in conjunction with airstrikes, ground operations, and joint missions with Malian and regional forces, such as the G5 Sahel Joint Force. This integration enabled a more cohesive approach to counterterrorism, where drones could provide real-time intelligence and immediate strike capabilities that enhanced the effectiveness of traditional military manoeuvres. For example, French drones are often deployed ahead of ground forces to conduct surveillance, identify militant positions, and guide troops safely through hazardous areas. During engagements, drones could deliver precise airstrikes on enemy positions, supporting ground troops and reducing the need for

[13] Bøås, M. and F. Strazzari, "Governance, Fragility and Insurgency in the Sahel: A Hybrid Political Order in the Making", *The International Spectator* 55, no. 4 (2020), 1–17.

[14] Kanig, C., "France Is Arming Its MQ-9 Reaper Drones", *European Security & Technology*, 20 December 2019, https://euro-sd.com/2019/12/articles/15672/france-is-arming-its-mq-9-reaper-drones/.

[15] Cooper, B., "The Sahel in West African History", *Oxford Research Encyclopedia of African History*, 26 September 2018.

[16] Ahmed, B. and A. Charlton, "France Calls Killing of Islamic State Leader Big Victory", *Associated Press News*, 16 September 2021, https://apnews.com/article/europe-africa-france-islamic-state-group-8c4f7b4b9e0b5f8d6d4b0e5e5f8e9f1a.

confrontations. After operations, french drones were used to assess damage, verify the elimination of targets, and ensure that militant groups could not quickly regroup.[17]

France's adoption of drone warfare in Mali reflects the broader global trend towards the use of technology in modern military conflicts. By leveraging drones, France was able to maintain a sustained and effective military presence in a challenging environment despite the logistical and operational difficulties inherent in the region. However, this strategy also raised questions about the long-term effectiveness of drone warfare, the potential for civilian harm, and the ethical implications of remote combat. As the conflict in Mali continues, these issues will remain central to the debate over the role of the outbreak of the Malian crisis in 2012. The country was on the verge of implosion because of weak institutions. The effect of decades of poor governance and the ineffective political and economic management of the state. The role of the corrupt ruling and governing elite subverting state institutions to serve their vested interests and regime survival, especially during the personalised rule of President Touré.[18] Both the Tuaregs and the diverse Islamist community were united in their fight against their common enemy, the government of Mali.[19] In a little more than two months, Mali's political regime was demolished based on bad governance. An armed rebellion led by the National Movement for the Liberation of Azawad (MNLA) launched on 17 January 2012 expelled the army from the north while a coup deposed President Amadou Toumani Touré (ATT) on 22 March 2012.[20] These two events ushered Mali into an unprecedented crisis, threatening regional stability. A group of prominent Malian political figures made an announcement condemning the coup as "a step backwards", including presidential candidate Ibrahim Boubacar Keita.[21] Political parties that condemned the coup included the Alliance for Democracy in Mali, the Union for the Republic and Democracy, the Patriotic Movement for Renewal, the Union for Democracy, the Union of Democratic Forces for Progress, the Solidarity and Progress Party, the Democracy and Justice Party and the Party for Democracy and Progress.[22] A US State Department Official also noted that Mali faced a "total cut off of foreign assistance" on which the nation heavily depends.[23]

---

[17] Vilmer, J. B. J., "A Perspective on France Proliferated Drones", *Center for a New American Security*, 2016.

[18] Morgan, A., "Coup Threatens to Plunge Mali Back into the Darkness of Dictatorship", *The Guardian*, 23 March 2012, www.theguardian.com/world/2012/mar/23/mali-coup-dictatorship (all websites cited in this publication were last accessed and verified on 1 December 2024).

[19] Ibid.

[20] Alozieuwa, S., "The March 22, 2012 Coup in Mali: A Setback for Democracy in Africa", *Democracy and Security* 9, no. 4 (September–December 2013), 383–397.

[21] Xinhua, "Malian Political Parties Condemn Coup d'Etat", *Xinhua News Agency*, 24 March 2012.

[22] Nossiter, A., Schmitt, E. and Mazzetti, M., "French Jets Strike Deep Inside Islamist Held Mali", *The New York Times*, 14 January 2013, www.nytimes.com/2013/01/14/world/africa/french-jets-strike-deep-inside-islamist-held-mali.html.

[23] Bondersholt, S.F. and C.K. Gyldenholm, "Conflict in Northern Mali: Tuareg Livelihood", *International Development Studies, Roskilde University* (2012), 12.

In the north, the Tuareg group which launched the rebellion, the National Movement for the Liberation of Azawad (Movement National deliberation de'l Azawad) MNLA, was silenced by an armed Islamist group, Ansar Dine (Ancar Eddine) led by Iyad Ag Ghali, a Tuareg Chief initially sidelined during the discussions which led to the creation of the MNLA. Having taken control of the north, Ansar Dine established a modus vivendi, if not a pact, with a range of armed actors, including former regime-backed Arab and Tuareg militias and, in particular, Al-Qaeda in the Islamic Maghreb (AQIM). The latter was responsible for kidnapping and killing many Westerners in Mali, Niger and Mauritania, attacks against the armies of the region and involvement in criminal trans-border trafficking; Northern Mali was becoming a haven for jihad fighters of all origins.

Mali, once regarded as a model of democratic progress in sub-Saharan Africa, has faced a severe crisis that has brought the country to the brink of dissolution. The northern regions of Mali, particularly areas like Timbuktu, Gao, and Kidal, have witnessed the rise of Islamist militant groups. These groups, including Ansar Dine, al-Qaeda in the Islamic Maghreb (AQIM), and the Movement for Oneness and Jihad in West Africa (MUJAO), have exploited local grievances and the collapse of state authority to consolidate their power. Their presence and strict enforcement of Sharia law have destabilised the region further, making any negotiated solution increasingly difficult. The political and institutional crisis in the capital, Bamako, has left the central government weak and ineffective. The 2012 military coup, which overthrew President Amadou Toumani Touré, created a power vacuum that has yet to be fully resolved. Successive governments have struggled to establish effective control over the country, and the military has been stretched thin, unable to restore order or protect the civilian population in the north. The growing influence of hardline Islamist factions has undermined efforts to negotiate a peaceful resolution to the conflict. The intransigence of these groups, combined with the weakened state apparatus in Bamako, means that any hope for a comprehensive peace deal is fading. As a result, Mali faces the grim possibility of further fragmentation, with the north increasingly disconnected from the central government and under the de facto control of Islamist militants. The international community has expressed concern, but the situation's complexity, with its deep-seated political, ethnic, and religious dimensions, makes it challenging to identify a clear path toward resolution.

ECOWAS (Economic Community of West African States) has played a significant role in attempting to mediate the crisis in Mali, but various challenges have marred its efforts.[24] ECOWAS, led by Burkina Faso's mediation team, was criticised for the opacity surrounding its diplomatic interventions. The lack of clear communication and transparency in the negotiation processes led to widespread scepticism and mistrust among Malians, particularly in Bamako. This opacity fuelled suspicions that the mediation efforts might not be fully aligned with the

[24] Okon, E.N., "ECOWAS Responses to the Conflict in Mali", *Conflict Studies Quarterly* 37, (2021), 36–53.

interests of the Malian people or that external agendas overly influenced them. Burkina Faso, under the leadership of President Blaise Compaoré at the time, took a central role in the mediation efforts. However, Compaoré's controversial political standing and Burkina Faso's regional ambitions cast a shadow over the mediation process. Many Malians and observers in the international community questioned whether Burkina Faso's involvement was truly neutral or whether it was driven by its interests, thereby undermining the perceived impartiality of ECOWAS's efforts. The diplomatic actions of ECOWAS, particularly under Burkina Faso's leadership, faced sharp criticism in the Malian capital and from various factions within the country and among Mali's regional and international partners. Many felt that the mediation process was out of touch with the realities on the ground, especially the complex dynamics between the northern insurgency and the political chaos in Bamako. This criticism further weakened the credibility of ECOWAS as a neutral and effective mediator.

As a result, while ECOWAS's involvement was crucial in preventing a complete collapse of the Malian state, its efforts were hindered by these perceived shortcomings. The lack of transparency and the controversial role of Burkina Faso led to a situation where ECOWAS's diplomatic actions were viewed with suspicion, limiting their effectiveness in achieving a sustainable solution to the crisis. "According to the UN High Commissioner for Refugees (UNHCR), 53,000 Malians have fled to Niger and 96,000 to Mauritania. UNHCR estimates that 174,000 Malians are internally displaced" because of recurrent conflict in Mali. However, pressure was mounting in support of an armed external intervention as specific security and political interests of foreign actors such as France, United States of America, European Union—neighbouring states and others—prevailed over those of the Malian population in both the north and south. In early January 2013, events overtook planning for an external intervention. Islamist fighters attacked towns in the central region of Mopti, perhaps hoping to capture critical infrastructure, like an airport in Sevare, which outside commanders hoped to use in staging their coming intervention. The Islamist advance evoked a swift response from Mali's former colonial ruler, France, which retains a vital political and security role in the Francophone African nation. Within days, French planes were bombing northern targets. In mid-January 2013, French and Malian soldiers recaptured GAO, Timbuktu, and Kidal. Troops from ECOWAS and Chad moved to join the French, hoping that after securing the north militarily, France would leave West African governments and Bamako to reassert political order.[25]

The battle for Mopti and the resulting French intervention underscored a critical lesson about the Malian crisis: the rapid, unpredictable tragedy can be explained with the benefit of hindsight from the way regime corruption weakened and angered the army to the level of MNLA's brutality created political opportunity for Islamists. However, the thread of continuity in events in Mali from 2012 to

[25] Thurston, A., "Mali: The Disintegration of a 'Model African Democracy'", *Stability Journal* 2, no. 1 (2013), 1–7.

2013 was how fragments and shifts render figures like Diarra powerful and raise hopes of progress, only to dash them soon after. In 2013, France started to deploy drones in Mali to fight Islamist groups. At this time, Malians welcomed the French intervention. The French government argued that the operation was essential for the security of France.[26] In Operation Serval, France used Class I drones, which are uncrewed aircraft with a mass lower than 150 kg, such as the French Survey Copters Cassidian DRACs and Harfang drones in Mali, to collect intelligence.[27] The French government stated that drones were a "unique tool" and "essential for modern operations".[28] Shortly after the start of Operation Serval, the then-French Minister of Defence Jean-Yves Le Drian explained that in the long term, fighter aircraft would be complemented with or even replaced by drones, as the deployment of drones was crucial in carrying out the French military operations.[29] In September 2013, an urgent order was placed through the US Foreign Military Sales programme for two unarmed MQ-9 Reaper drones, which would be deployed in Mali in January 2014.[30] When the military programme for 2014–2019 was adopted, the urgent need for drones to collect intelligence was again highlighted. It stated that France would acquire twelve Class III drones (weighing more than 600 kilograms) before 2019 and possess 30 Class II drones with a higher level of risk by 2025. The Class III drones, stationed at the US military base in Niamey in Niger, were extensively used to collect in-theatre intelligence for military operations, including acquiring enemy targets on the ground and helping other armed aircraft with targeting using laser sensors.[31]

In 2014, for example, ten suspected Islamist fighters were killed in Mali after they had been spotted by a French drone.[32] Their growing use has been driven by both their use in counterterrorism operations for targeted killings and the need for more situational awareness and information in regular military operations through

---

[26] France24, "French Military to Extend Mali 'Counterterrorism' Operations into Sahel", *France 24*, 13 July 2014, www.france24.com/en/20140713-france-military-extend-mali-counter-terrorism-operations-sahel.

[27] Le Monde, "Hollande: l'opération au Mali «n'a pas d'autre but que la lutte contre le terrorisme", *Le Monde*, 12 January 2013, www.lemonde.fr/afrique/article/2013/01/12/hollande-l-operation-au-mali-n-a-pas-d-autre-but-que-la-lutte-contre-le-terrorisme_1816493_3212.html.

[28] Ministère de la Défense, "French White Paper: Defence and National Security", 2013.

[29] Le Drian, J-Y, "Pourquoi l'armée française a un besoin urgent de drones", *Les Echos*, 31 May 2013, https://www.lesechos.fr/2013/05/pourquoi-larmee-francaise-a-un-besoin-urgent-de-drones-1097785.

[30] Stevenson, B., "France Orders Third Reaper System", *Flight Global*, 17 December 2015, www.flightglobal.com.

[31] French Harfang UAV logs 2000 flight hours in Mali', (Shephard Media, 5 September 2013, https://www.shephardmedia.com/news/uv-online/french-air-force-harfang-logs-2000-flight-hours-ma/).

[32] Chicago Tribune, "French, Mali Troops Kill 10 Suspected Islamist Militants", 6 March 2014, www.chicagotribune.com/news/ct-xpm-2014-03-06-sns-rt-us-mali-fighting-20140306-story.html.

a range of cameras and sensors fitted to the drones.[33] Since 2016, the Malian population has organised multiple protests against foreign forces operating in the country.[34] The Malian citizenry frequently protested France's presence in Mali to jettison France from Mali.[35] The population accused the French army of going after innocent people without sharing information on why they did so.[36]

Moreover, neither the French forces nor the Malian government gave the population any information about the various French military operations. At the time, Major General Patrick Bréthous told Malian journalist Baba Ahmed, "We are not here to highlight the results of our operations, but rather to ensure that the armed terrorist groups no longer have any sanctuary".[37] He argued that the Malian authorities should decide whether to provide information about the military operations. However, the Malian authorities have harassed journalists who cover security issues. In addition, criticising the army can result in arrests on charges of contravening standards and undermining troop morale. The opaqueness of the French intervention led to more unrest in the Malian state. In 2017, hundreds of Malians protested at the French embassy in the capital, saying that France was not impartial in the conflict between the Malian government and the armed groups.[38] Some protesters were not calling on France to leave the country but wanted to know what the French military was doing in the region, as they suspected France was attempting to gain control of Mali's natural resources.[39] In the same year, locals protested in Kidal in northeast Mali and attacked French forces on patrol with stones.[40]

Again, in 2018, Malians protested at the French embassy in Bamako.[41] The anti-French sentiment among the Malian population rose even further in 2019

[33] Bachmann, S. D., "Targeted Killings: Contemporary Challenges, Risks and Opportunities", *Journal of Conflict and Security Law* 18, no. 2 (2013), 259–288.

[34] Ahmed, B., "Gal Patrick Bréthous (Barkhane): au Mali, les terroristes fuient les forces internationales", *Jeune Afrique*, 29 July 2016, https://www.jeuneafrique.com/345701/politique/gal-patrick-brethous-barkhane-mali-terroristes-fuient-forces-internationales/.

[35] Carayol, R., "Au Sahel, le spectre de la menace fantôme", Mediapart, 6 September 2020. https://www.mediapart.fr/journal/international/060920/au-sahel-le-spectre-de-la-menace-fantome.

[36] Ahmedou, M. and Bayle, T., "Mali: A Kidal, l'armée française au pied du mur", *Mond Afrique*, 20 April 2016, https://mondafrique.com/mali-a-kidal-larmee-francaise-pied-mur/.

[37] Ahmed, B., "Gal Patrick Bréthous (Barkhane): au Mali, «les terroristes fuient les forces internationales»", *Jeune Afrique*, 29 July 2016, https://www.jeuneafrique.com/345701/politique/gal-patrick-brethous-barkhane-mali-terroristes-fuient-forces-internationales/.

[38] BBC, "Why France is Focused on Fighting Jihadists in Mali", *BBC*, 27 November 2019, www.bbc.com/news/world-africa-50558972.

[39] Belsoeur, R. and Tagnan, A., "Au Sahel, la colère sourde des populations contre les troupes françaises", *Slate*, 17 August 2017, http://www.slate.fr/story/149784/burkina-sahel-terrorisme-soldats-francais.

[40] Richard, J., "CTS, Counterterrorism and Non-Violence", *Critical Studies on Terrorism* 10, no. 2 (2017), 357–369.

[41] Ahmed, B., "Au Mali, la multiplication des attaques renforce le sentiment anti-français", *Jeune Afrique*, 24 October 2019, www.jeuneafrique.com/842682/politique/au-mali-la-multiplication-des-attaques-renforce-le-sentiment-anti-francais/.

after at least 41 Malian soldiers were killed in an attack on the military camp of Boulkessi.[42] A general suspicion grew among the Malian population that France was assisting the jihadists in Mali, as France was unable to halt the attacks on the population despite its sophisticated weapons and drones. This conspiracy theory circulated widely on social networks in Mali.[43]

France carried out its drone strike in Mali in December 2019, only two days after the army finished testing drones for armed operations.[44] In the operation, in which helicopters and ground troops were deployed as well, 40 "terrorists" were "neutralised". Ten days later, nine terrorists were "put out of action" with the aid of a combat helicopter and armed Reaper drone.[45] In January 2020, 35 militants and 23 motorcycles were "neutralised" by combat helicopters and an armed drone. A month later, the French government announced once again that it had "neutralised" 50 militants and destroyed 30 motorcycles and two pickup trucks in Mali with the help of a Reaper drone, Mirage 2000 airstrikes and combat helicopters.[46]

Moreover, press releases let the public believe there have been almost no civilian casualties in the French military operations. However, it is tough for the French military to know who the "terrorists" are. A French soldier who fought in Mali in 2019 said, "The jihadists hide among the populations. Every man is a potential suspect".[47] Thus, it is widely speculated that the deployment of armed drones in the Sahel has resulted in civilian deaths. In February 2020, France stated in a press release that it had "neutralised" 20 terrorists in a military operation in Burkina Faso with various aircraft, including a Reaper drone.[48] French news portal Media Part has collected testimonials that indicate that during that military operation, several civilians were killed by a French drone.[49] Furthermore, several sources have

---

[42] Ahmed, "Au Mali, la multiplication des attaques renforce le sentiment anti-français", 24 October 2019.

[43] Munshi, N., "How France Lost Mali: Failure to Quell Jihadist Threat Opens Door to Russia", *Financial Times*, 22 December 2021, www.ft.com/content/5c4e6efb-2b1b-4f58-b064-4d7f7f0fa705.

[44] France24, "French Military to Extend Mali 'Counterterrorism' Operations into Sahel", *France 24*, 13 July 2014, www.france24.com/en/20140713-france-military-extend-mali-counter-terrorism-operations-sahel.

[45] Ministère des Armées, "Point de situation des opérations du 3 au 9 janvier", 9 January 2020, https://www.defense.gouv.fr/operations/points-de-situation/point-de-situation-des-operations-du-3-au-9-janvier.

[46] Ministère des Armées, "Point de situation des opérations du 3 au 9 janvier", 9 January 2020; SLD, "Operation Barkhane Update, April 2020", Second Line Defence, 21 April 2020, https://sldinfo.com/2020/04/operation-barkhane-update-april-2020/.

[47] Carayol, "Au Sahel, le spectre de la menace fantôme", *Mediapart*, 6 September 2020.

[48] Ministère des Armées, "Point de situation des opérations du 3 au 9 janvier", 9 January 2020; SLD, "Operation Barkhane Update, April 2020", Second Line Defence, 21 April 2020.

[49] Carayol, "Au Sahel, le spectre de la menace fantôme", *Mediapart*, 6 September 2020.

reported that civilians were killed by French drone strikes in Gourma.[50] However, these strikes are not verified as there are no independent watchdogs in the area.[51] In 2020, the deterioration of the conflict, civilian deaths, and the start of armed drone operations by France further increased the Malian population's opposition to the French military intervention.[52] Anti-France demonstrations continued, and so did the spread of conspiracy theories.

Moreover, at least some of the local population has become too terrified to talk to the French army.[53] When French soldiers talk to locals to collect intelligence for their military operations and then move on, the people who helped the French military can then be attacked or executed by non-state groups. The reason is farfetched when local civilians provide intelligence to French forces; they risk being perceived as collaborators by militant groups. These groups may retaliate by attacking or executing locals accused of aiding foreign forces, causing villagers to feel insecure despite the presence of international military operations. Armed drone operations, while precise, can create a sense of anxiety among villagers. Drones rely heavily on intelligence, and any interaction with foreign soldiers might make civilians fear being mistakenly identified as insurgents or collaborators. Even if they aren't directly involved, the mere presence of foreign military personnel in their villages could lead to suspicion and possible drone strikes. For some Malian communities, the perceived or real risks of being associated with foreign military forces, whether through intelligence sharing or proximity during drone operations, can erode trust in those forces. Rather than feeling protected, they might feel caught between foreign forces and insurgent groups, both of which pose threats to their safety.[54]

4 Impact and Effectiveness of Drone Warfare in Mali

France's adoption of drone warfare in Mali has yielded several tactical successes in its ongoing efforts to combat extremist groups. Drones have significantly enhanced the French military's ability to conduct surveillance and reconnaissance across

[50] Freudenthal, E., Huon, P., Nsaibia, H., van der Weide, Y. and Bolly, M., "The New Humanitarian, "Uncovering the Civilian toll of France anti-Jihadist War in Mali", *The New Humanitarian*, 16 June 2021, www.thenewhumanitarian.org/investigations/2021/6/16/uncovering-the-civilian-toll-of-france-anti-jihadist-war-in-Mali.

[51] Carayol, "Au Sahel, le spectre de la menace fantôme", *Mediapart*, 6 September 2020.

[52] Freudenthal, et al., "The New Humanitarian, "Uncovering the Civilian toll of France anti-Jihadist War in Mali", *The New Humanitarian*, 16 June 2021.

[53] Carayol, "Au Sahel, le spectre de la menace fantôme", Mediapart, 6 September 2020.

[54] Guifford, J., "Security Collapse in Mali and Burkina Faso: What Implications?—Anticipating the Crisis Through the Eyes of Jihadists", *Institut Montaigne*, 11 January 2023, www.institutmontaigne.org/en/analysis/security-collapse-mali-and-burkina-faso-what-implications; Tapily, M., "As Army Operations Ramp Up in Mali, Rebel Groups Impose 'Suffocating' Blockades", *The New Humanitarian*, 12 March 2024, www.thenewhumanitarian.org.

the vast, challenging terrain of the Sahel.[55] The real-time intelligence provided by drones has been crucial in identifying and tracking the movements of militant groups such as Al-Qaeda in the Islamic Maghreb (AQIM) and the Islamic State in the Greater Sahara (ISGS). This intelligence has enabled French forces to disrupt militant operations, ambushes, and supply lines more effectively. The precision capabilities of armed drones have also allowed France to carry out targeted strikes on high-value targets with minimal collateral damage.[56] These strikes have resulted in the elimination of key militant leaders and the destruction of critical assets, which has disrupted the command structures and operational capabilities of these groups.[57] For example, the elimination of high-profile figures within militant organizations has led to temporary disarray within their ranks, reducing their ability to coordinate attacks and spread their influence across the region.[58] Despite these tactical successes, the use of drone warfare in Mali has faced several limitations and challenges. One of the primary challenges is the difficulty of gathering accurate and actionable intelligence in a complex and fluid environment.[59] Militant groups in the Sahel often operate in small, mobile units, blending in with local populations or retreating to remote areas where they are difficult to track.[60] This makes it challenging to identify legitimate targets and avoid civilian casualties, which can undermine the overall effectiveness of drone operations.

The reliance on drones also raises concerns about the sustainability of such a strategy. While drones offer significant tactical advantages, they cannot address the underlying political and socio-economic issues that fuel instability in Mali.[61] The effectiveness of drone strikes in the short term may not translate into long-term stability if these root causes are not addressed. Moreover, the limited number of drones available and the logistical challenges of operating them in such a vast region mean that drone coverage is often insufficient to monitor all areas where

---

[55] Lushenko, P., Raman, S. and Kreps, S., "Multilateralism and Public Support for Drone Strikes", *Research and Politics* (April–June 2022), 1–9.

[56] Thomas, G. N., "In Africa, U.S. Military Struggles to Help Allies Counter Terror", *The New York Times*, 24 October 2019, www.nytimes.com/2019/10/24/world/africa/mali-africa-us-france-counterterrorism.html.

[57] Callimachi, R. and E. Schmitt, "French Forces Kill Top Extremist Leader in West Africa", *The New York Times*, 24 September 2020, www.nytimes.com/2020/09/24/world/africa/mali-france-kill-qaeda.html.

[58] Jean-Hervé, J., "Mali: From Sanctuary to Battlefield", *International Crisis Group*, August 1, 2019. https://www.crisisgroup.org/africa/sahel/mali.

[59] Lewis, D., "Remote Control Warfare:Drone Strikes and the Challenges of Asymmetry in the Sahel", Reuters, 5 May 2021, https://www.reuters.com/article/sahel-drone-strikes-idUKSKBN2CU08J.

[60] Jacob. Z., "Militant Mobility and the Challenges of Counterterrorism in the Sahel". *Jamestown Foundation*, 15 June 2020, https://jamestown.org/program/militant-mobility-and-the-challenges-of-counterterrorism-in-the-sahel/.

[61] Alexander, T., "A Closer Look at French Counterterrorism in the Sahel", *War on the Rocks*, 1 February 2021, https://warontherocks.com/2021/02/a-closer-look-at-french-counterterrorism-in-the-sahel/.

militants might operate. The impact of drone warfare on the local population in Mali has been mixed. On one hand, drone operations have contributed to a reduction in large-scale militant attacks in some areas, which has brought a measure of security and stability to regions previously under threat.[62] On the other hand, the presence of drones and the fear of being targeted have led to anxiety and resentment among civilians, particularly when strikes result in civilian casualties or when there is a perceived lack of accountability for mistakes.

The use of drones has also complicated relations between the French military and the local population. While some communities view the drones as a necessary tool in the fight against extremism, others see them as an extension of foreign intervention that exacerbates the conflict.[63] This perception can fuel anti-French sentiment and potentially drive some individuals towards supporting militant groups, viewing them as defenders against external forces. The effectiveness of drone warfare in Mali also has broader strategic implications for France's military engagement in the Sahel. Drones have allowed France to maintain a sustained military presence with relatively low risk to its personnel, which is politically advantageous given the growing domestic opposition to prolonged military interventions. However, the reliance on drones may also create a false sense of security, leading to an overemphasis on military solutions at the expense of political and diplomatic efforts to resolve the conflict.

Moreover, the use of drones has implications for international norms and the rules of engagement in counterterrorism operations. The expansion of drone warfare by France sets a precedent for other countries, potentially lowering the threshold for the use of force and complicating efforts to establish global standards for the use of unmanned systems in conflict zones. With increasingly more use of military drones in warfare, an increasing number of experts have voiced serious criticisms of drone deployment. First, experts explain that drones can make it easier to resort to the use of lethal force and may contribute to conflict escalation.[64] Drones have relatively low operating costs and can be operated without direct risk to the pilot. This could make the option of targeted killings by armed drones or with the support of unarmed drones more attractive than, for example, detaining or arresting suspects. Secondly, drones are hailed for their precision strikes or ability to support precision strikes in combat operations. Drone strikes can indeed be more precise compared with using 'dumb 'munitions. However, human error can

[62] Fuhrmann, M. and J. Schwartz, "Drone Attacks Can Result in Blowback Among Civilian Populations", *The Washington Post*, 18 August 2022, www.washingtonpost.com/politics/2022/08/18/drone-attacks-can-result-blowback-among-civilian-populations/.

[63] DeVore, M.R., "Strategic satisficing: Civil-military Relations and French intervention in Africa", *European Journal of International Security* 4, no. 2 (2019), 163–189.

[64] Callamard, A., "Use of Armed Drones for Targeted Killings: Report of the Special Rapporteur on Extrajudicial, Summary or Arbitrary Executions", United Nations, 2020, www.ohchr.org/en/documents/thematic-reports/ahrc4483-use-armed-drones-targeted-killings-report-special-rapporteur.

result in civilian injuries or deaths when armed or unarmed drones are used to support attacks.[65]

The targets of attacks are derived from networks of informants and data collected by drones. Different teams across various agencies and countries are often involved in analysing the intelligence and visuals from drones, which can increase the risk of inaccurate use of the technology.[66] Third, in so-called 'signature strikes' the targets are selected based on certain characteristics associated with the combatants, such as carrying weapons, and metadata, such as gender and age. Yet the visual distinctions between combatants and civilians are often at the very least ambiguous, if not non-existent.[67] Even though drones can strike more precisely than 'dumb' munitions, this does not mean that they do not create civilian casualties, nor that they are less prone to human error, as the teams operating them have to understand and interpret the incoming data from sensors and the underlying algorithms while working together, and to decide who is a suspect and who is not.

France considers itself a party to the armed conflict in Mali at the invitation of the Malian government and depends on the consent of neighbouring countries in expanding its counterterrorism campaign in the Sahel region. It will interest you under international law; a state's consent may preclude the wrongfulness of acts that would otherwise be contrary to international law. As part of its jus ad bellum framework, Article 2 (4) of the UN Charter prohibits the use of force "against the territorial integrity or political independence of any state (Article 2 of UN Charter). A foreign power deploys military force in another state's territory, but within the validly given consent of the host state. This would constitute a violation of the UN charter. This approach to civil war and counter insurgency will not work well when it comes to fighting (especially) extremist transnational organized and well supplied groups like al Qaeda in the Islamic Maghreb (AQIM). In fact, French military forces were filling the military vacuum as a result of the lack in the capacities and capabilities of Mali African forces.[68] This has a clearly evolving pattern of counterterrorism operation in Mali, with France forces using technologically advanced weaponry including UAVs and Mali troops serving on the ground as garrison forces for towns and villages.

Finally, looking at the domestic and organizational consequences of security strategies based on the use of UAVs, there is strong evidence that civilians in Mali have been injured and killed by drone strikes.[69] Drone users such as France, Russia, United Kingdom, Iran, United States of America may have mistakenly identified these civilians as armed militants or simply have been applying international law obligations inadequately. Worrisomely, governments often refuse to

---

65 Ibid.

66 Ibid.

67 Espinoza, "State Terrorism: Orientalism and the Drone Programme", 2018.

68 Shurkin, M., Pezard, S. and Zimmerman, R., "Mali's Next Battle: Improving Counterterrorism Capabilities", *RAND Corporation*, 20 April 2017, https://www.rand.org/pubs/research_reports/RR1241.html.

69 Carayol, "Au Sahel, le spectre de la menace fantôme", *Mediapart*, 6 September 2020.

admit unlawful attacks, and thereby try to avoid taking responsibility for the victims of these attacks. Victims and relatives of victims of unlawful drone attacks have a right to reparations and access to information about alleged violations and investigations into such violations.[70] The presence of foreign forces and their use of drones have led to protests in Mali Burkina Faso, and Niger in the region.[71] Civilians were afraid the use of drones would kill or injure more innocent civilians.

5 Conclusion

The utilisation of drones in Mali illustrates the significant role that advanced technology now plays in modern warfare and humanitarian efforts. Drones have proven to be invaluable assets for the French military, providing enhanced surveillance, intelligence, reconnaissance, and precision strike capabilities. These tools have allowed military forces to conduct more efficient operations, reduce risks to personnel, and improve the effectiveness of counterterrorism efforts in a challenging and vast terrain like the Sahel.

However, the growing reliance on drone warfare in Mali has also raised important ethical and legal concerns. Civilian casualties, sovereignty violations, and the potential for misuse by non-state actors highlight the need for clear and robust regulatory frameworks. The use of drones demands strict adherence to international humanitarian laws and human rights principles to ensure accountability and transparency in military operations.

As the conflict in Mali continues, it is essential for both national and international stakeholders to consider the broader implications of drone warfare. Technological superiority alone cannot address the deep-seated socio-political and economic issues that fuel instability in the region. Therefore, while drones offer significant tactical advantages, long-term peace and stability in Mali will require comprehensive strategies that combine military, political, and diplomatic efforts.

Moving forward, the responsible and ethical deployment of drone technology will be crucial in promoting security and stability in Mali and the broader Sahel region. By fostering collaboration among military forces, humanitarian organizations, and local authorities, stakeholders can leverage the benefits of drone technology while mitigating its risks. Ultimately, the success of such efforts will depend on balancing military efficiency with the protection of civilians and the promotion of sustainable development.

---

[70] Heyns, C. and Akande, D., "The International Law Framework Regulating the Use of Armed Drones", *International & Comparative Law Quarterly* 65, no. 4 (2016), 791–827.

[71] Okpaleke, F, "Drones Are a Crucial Counterinsurgency Tool in the Sahel", *London School of Economics Blog*, 2023. https://blogs.lse.ac.uk/africaatlse/2023/03/15/drones-are-a-crucial-counterinsurgency-tool-in-the-sahel/.

Arinze Ngwube is a lecturer in the Department of Political Science at Federal University Oye-Ekiti, Ekiti State, Nigeria. A seasoned commentator on regional conflicts and strategic trends, he frequently contributes to television and radio discussions on international affairs and security dynamics. His research interests include peacebuilding, African politics, conflict resolution, the politics of natural resources, and strategic studies. He holds a B.Sc. in Political Science from Babcock University and an M.A. in International Relations from London Metropolitan University. He is currently pursuing a Ph.D. in Political Science at the University of Ilorin. Ngwube is an active member of several organisations, including the International Political Science Association (IPSA), the Nigerian Political Science Association (NPSA), Chatham House, the Nigerian Institute of Management (NIM), and the Educational Research Network for West and Central Africa (ERNWCA).

Exploring the Securitisation of Drones in the South African Military

Kyle John Bester and Danille Elize Arendse

Abstract

Unmanned aerial vehicles (UAV's) have become increasingly relevant in the African continent due to its capability to advance issues related to peace and security. However, the proliferation of UAV's whether it is armed or unarmed is becoming a security challenge for nation states because of the concerns associated with privacy, public safety and the utilisation of drones by non-state actors for malicious purposes. The accessibility, affordability and capability to be weaponised positions UAVs as a threat to maintaining security. In contrast, the use of technological advancement such as UAVs contribute to the outcome of the African Union Agenda 2063 by establishing a more peaceful and security orientated Africa. Nevertheless, in the South African armed forces context, contextual challenges such as continued budgetary constraints, and a limited integration of technology may hinder this advancement of Agenda 2063. This paper conceptually explores UAVs as a 21st century threat through the theoretical lens of Securitisation Theory (ST) within the South African armed forces context. ST allows for the exploration of threats emerging in a social context by highlighting the security actors involved and the social value placed on the meaning of threats. The focus on UAVs with a South African armed forces context provides a nuanced perspective on security, particularly through the speech act. The paper further explores the contextual and security challenges that UAVs pose to national security as well as the various attempts by securitising actors to elevate UAVs as an emerging threat in the SA armed forces context. The author argues that exploring the securitisation of UAVs within the

K. J. Bester (✉)
University of South Africa (Unisa), Johannesburg, South Africa
e-mail: bestekj@unisa.ac.za

D. E. Arendse
Department of Psychology, University of Pretoria, Pretoria, South Africa
e-mail: u14458200@up.ac.za

W. S. Coetzee and D. Putter (eds.), *Drones in the African Battlespaces*, Southern Space Studies, https://doi.org/10.1007/978-3-031-87147-4_6

South African armed forces context allows for a critical discourse of its effectiveness within the maintenance of security in South Africa. This paper aims to contribute to the global debate on UAVs and add to the dearth of literature on the securitisation of drones in the South African armed forces context.

1 Introduction

War is changing and the use of soldiers, particularly occupying traditional roles, is also shifting. Drones shape the nature of warfare and minimising the risk that war has on soldiers, specifically the physical risk impact thereof. The aim of drones in the military is to increase operational functionality while lowering the possibility of death or injury for soldiers on the ground. In some cases, the use of drones has been compared to that of video games, allowing the soldiers to take a distanced approach to warfare. This should however not be an indication of risk, as these soldiers are still at risk for traumas, albeit a different kind of trauma. Although there is an increase in drone usage in the military, it is still considered an emerging field.[1] The use of drones by militaries globally has advanced at a rapid pace, and more armed forces are considering UAVs as a tool for advancing their force capabilities. The availability of drones varies in size and capabilities and can be accessed. The variation of drones also possesses a variety of functionalities, such as video and recording capabilities that are particularly suitable for collecting data. The technological development of these capabilities makes using drones useful for agencies within a security cluster to gather intelligence through surveillance and tactical activities. Using drones is also not void of the ethical and political challenges globally. The use and development of drones in the European context have mainly been centered within the civilian realm. However, the military has significantly benefitted from this development, whereby force capabilities are increased. In the African context, the expansion and accessibility of drone technology are increasing exponentially on the continent, which also has to do with China's technological investment. Considering the expansion, these civilian-grade commercial drones can be weaponised.[2]

The United States of America (USA) has been foremost in the advancement and operation of drones. Their operationalizing of drones in the US military has been for both investigative purposes using cameras and for weapons since they are able to transport missiles. The US military uses drones for both domestic and international purposes, thus extending their reach when embarking on different

[1] Molenkamp, A., Weerdesteijn, M. and Smeulers. A., "A Playstation Mentality to Killing? Adverse Psychological Consequences in Drone Pilots and the Stigmatization Thereof in the Military", *Armed Forces and Society* (2024), 1–20, https://doi.org/10.1177/0095327X24123622.

[2] African Defence Forum., "Small, Cheap and Dangerous", *African Defence Forum*, 21 October 2024, https://adf-magazine.com/2024/10/small-cheap-and-dangerous/.

missions. Drone pilots were instrumental in the location and information profile of terrorists for the US military while they were involved in missions in the Middle East. They have also actively used drone pilots in the US military to engage in targeted missions in which terrorists are killed using drones. This operation is however not only upon the drone pilot but relies on several aspects of the drone mission before a suspected terrorist is killed. The drone mission relies on an intelligence coordinator, who receives information from the intelligence analyst. These intelligence analysts are responsible for analysing the visual material received by the drones. Thus, once sufficient information is received and a decision is made, the drone pilot then becomes activated for his targeted mission for the terrorist.[3]

Over the years, there have been various drone attacks, each varying in impact and extending the use of drones across different contexts. Some of the most notable examples are namely: the spraying of gas on Tokyo residents in 1994, the identification of nine drones by the Colombian army in 2002, a drone attack in Syria in 2014, drones used to drop grenades in a Ukrainian arms depot and the attempted assassination of Venezuelan president by drones in 2018.[4] It is also worth noting that Syria was the site of the first drone attack by a suspected terrorist using a drone and subsequently, the Syrian civil war has been the greatest example of drone warfare with a noted 32 drones in the sky.[5]

The use of drones has been cautioned by Majeed et al., of its potential impact on a variety of sectors within civil society.[6] Cameron draws on the notion that drones are the new Internet of Things (IoT) and that they primarily serve to improve the lives of human functioning through enhanced battery life and assisting with cognitive functioning. Whilst the focus of Cameron was on battery life, it may also pose some security concerns related to frequency and time nefarious actors may engage in using drone technology to execute their activities.[7] Majeed et al., describe a drone as mobile and lightweight where it can aid the security clusters in monitoring criminal activity. However, drones may also be equipped with harmful objects to execute certain criminal activities. In addition, drones may also exert damage to civilian life and property and may even be used as a tool for extremist activities. Therefore, drones may compromise the survival of human life and threaten existing security measures and laws that govern the survival of the state.

---

[3] Molenkamp, et al., "A Playstation Mentality to Killing? Adverse Psychological Consequences in Drone Pilots and the Stigmatization Thereof in the Military", 2024.

[4] Chávez, K. and Swed, O., "The Proliferation of Drones to Violent Nonstate Actors", *Defence Studies* 21, no. 1 (2021), 1–24, https://doi.org/10.1080/14702436.2020.1848426.

[5] Chávez and Swed, "The Proliferation of Drones to Violent Nonstate Actors", 2021.

[6] Majeed, R., Abdullah, N.A., Mushtaq, M.F. and Kazmi, R., "Drone Security: Issues and Challenges", *International Journal of Advanced Computer Science and Applications* 12, no. 5 (2021), 720–729, https://doi.org/10.14569/IJACSA.2021.0120584.

[7] Cameron, L., "Drones as the New 'Flying IoT'", www.computer.org/publications/tech-news/research/flying-iot-toward-low-power-vision-sky.

Majeed et al., and Finn Wright suggest that the notion of security may not necessarily lend itself to an outcome of protection.[8] Instead, the presence and use of drones present nation-states with a conundrum where the double-edged sword analogy applies, since drones may pose significant concerns related to human survival (physical and economic implications such as financial loss). Furthermore, Allworth also alludes to the threat capability of the UAVs by pointing out that nefarious actors may also use Artificial Intelligence (AI) to 3D print drones.[9] Thus, it should be questioned whether 4IR (Fourth Industrial Revolution) advances the idea of ensuring National Security or poses security challenges that may restrict technology being fully integrated within the security domain.

The technological advancement in drone technology has opened new opportunities in various fields and brought new security challenges.[10] Drone technology represents a groundbreaking innovation that offers a myriad of advantages and benefits for human beings. Drones are instrumental in facilitating a wide range of day-to-day activities, as well as in military operations and the monitoring of weather patterns. Nevertheless, these remarkable capabilities also give rise to significant concerns regarding privacy and safety. It is imperative that these concerns, particularly those pertaining to security and privacy breaches, are effectively and comprehensively addressed.[11] Furthermore, drones offer a wide array of applications. However, the security concerns regarding this use may stretch over to other domains, such as cyberspace.[12] Bester and Arendse notes that cyber threats are pervasive and pose significant security risks for users.[13] In this case, the use of drones may facilitate threats and attacks in a physical or digital capacity. Drones can present security risks in the form of physical or cyber-attacks.[14] To address these risks, it's crucial to impose restrictions on drone usage in civilian areas and private properties. The misuse of drones is on the rise, giving rise to numerous challenges for nation-states and their inhabitants. Operators of drones often exploit Bluetooth or Wi-Fi channels to navigate their devices in restricted zones, potentially resulting in significant financial loss, thus impacting the economy of a sector. In addition, drones may also be used to compromise Wi-Fi networks and intercept

[8] Majeed, et al., "Drone Security: Issues and Challenges", 2021; Finn, R.L. and Wright, D., "Unmanned Aircraft Systems: Surveillance, Ethics and Privacy in Civil Applications", *Computer Law & Security Review* 28, no. 2 (2012), 184–194, https://doi.org/10.1016/j.clsr.2012.01.005.

[9] Allworth, E.M., "The South African Air Force (SAAF), Unmanned Aircraft Systems and National Security: An Exploratory Study" (Unpublished Masters Dissertation, Stellenbosch University, 2018).

[10] Majeed, et al., "Drone Security: Issues and Challenges", 2021.

[11] Ibid.

[12] Ibid.

[13] Bester, K.J. and Arendse, D.E., "Measuring Cybersecurity Awareness: Exploring the Reliability and Factor Structure of the Cyber Security Questionnaire in a South African Military Sample", *South African Journal of Military Studies* 52, no. 1 (2024), https://doi.org/10.5787/52-1-1445.

[14] Majeed, et al., "Drone Security: Issues and Challenges", 2021.

Bluetooth connections, leading to many privacy and security concerns for individuals. This ability to intrude on secure networks may only exacerbate the global and domestic attempts to reduce the cyber threat rate.[15]

Expanding on the discussion, one must highlight the concept of security within the broader argument of unregulated drone use as an emerging threat. Security is multifaceted, with varying interpretations across different academic disciplines, including law, geography, criminology, psychology, sociology, politics, and international relations.[16] For the purpose of this chapter, the focus will be on examining security through the lenses of psychology, sociology, and international relations, emphasising the role of human factors in understanding security threats. Bourbeau et al., advocate a holistic approach to studying security, drawing from diverse disciplinary perspectives. Romaniuk challenges the conventional notion of security as solely tied to politics and national security, proposing that security is a socially constructed and subjective concept, providing insights into the societal challenges originating from nation-states, civil society, armed forces, politics, economics, and the environment. Romaniuk also highlights that changes in security threats within these domains can have far-reaching impacts on individuals and society as a whole.[17] Additionally, Philipsen points out that the involvement of new security actors may further expand the understanding of what constitutes security threats.[18]

The concept of sovereignty is worth considering as sovereignty becomes undermined using technologies such as drones, since they can permeate borders. As a result, "drone usage has been justified by the purported inability or unwillingness of some states to maintain their territorial integrity and border security".[19] This places the idea of sovereignty and border control in a precarious state as drones manoeuvre into spaces and territories. This brings the notion of terrorism as a threat that can emerge anywhere and everywhere, and countries need to be ready to protect themselves.[20] This also implies that borders may become porous if some countries deem it fit to engage in warfare beyond their own borders as part of ensuring international peace. Drones thus become an ideal vehicle through

[15] Bester and Arendse, "Measuring Cybersecurity Awareness: Exploring the Reliability and Factor Structure of the Cyber Security Questionnaire in a South African Military Sample", 2024.

[16] Bourbeau, P., Balzacq, T. and Cavelty, M.D., "International Relations: Celebrating Eclectic Dynamism in Security Studies", in *Security: Dialogue Across Disciplines*, ed. P. Bourbeau (Cambridge University Press, 2015), 111–136.

[17] Romaniuk, S.N., "Copenhagen School" in *The SAGE Encyclopaedia of Surveillance, Security and Privacy* (SAGE, 2018), http://dx.doi.org/10.4135/9781483359922.n95.

[18] Philipsen, L., "Performative Securitization: From Conditions of Success to Conditions of Possibility", *Journal of International Relations and Development* 23 (2018), 139–163, https://doi.org/10.1057/s41268-018-0130-8.

[19] Johnson, B., "Coded Conflict: Algorithmic and Drone Warfare in US Security Strategy", *Journal of Military and Strategic Studies* 18, no. 4 (2018), 49.

[20] Ibid.

which different countries can navigate issues related to sovereignty, security and peace.[21]

In South Africa, the commercial use of drones in civil society is subject to specific guidelines and regulations, limiting their capabilities and reach. Private drone users in South Africa are considered recreational drone operators. The laws and guidelines are enforced by the civil aviation authority, ensuring a clear distinction between drones used for recreational purposes and those equipped for military use. Although regulations are in place, the South African Aviation Authority may not have the full capacity to strictly enforce them. According to Majeed et al., the use of drones in private areas without the consent of the property owner is not permitted. Furthermore, to fly drones in private areas, drone operators are expected to acquire permission certification from the property owner. Failing to adhere to and comply with guidelines and legislation may result in repercussions.[22]

The position of the SANDF in the SA context has taken an approach where there is active collaboration between the private and armed forces sectors. It has shown the continuous stance of the advancement of drones in the African region. The private entities at the forefront of developing drones, such as DENEL and Milkor, have formed close relationships with the SANDF, thus making it possible for drone concepts to be developed that range from hand-launched capabilities to larger drones. The exploration of securitisation theory and its relevance to drones in the SA armed forces context remains largely unexplored. This could be due to research on drones in the social science context of being in their infancy. Furthermore, Securitisation Theory (ST) explains how ordinary security issues are elevated into the realm of politics. This chapter takes the positionality of ST to understand the relevance of drones in the SANDF context. Furthermore, this chapter aims to explore using drones as a weaponised response to security challenges within the SANDF. Ultimately, this chapter contributes to research on new conceptualisations of security by focusing on drones and using ST as a theoretical guide in positioning how this new and emerging threat is framed. This chapter highlights the attempts to introduce drones as a response to emerging threats in the SA armed forces context.

2 Significance of Securitisation Theory

Securitisation Theory (ST) focuses on elements associated with war and conflict, particularly when considering the successful securitization process. This level of success does not necessarily imply that all aspects of the theory were fully addressed; it signifies the resolution of addressing existential threats through

[21] Ibid.

[22] Majeed, et al., "Drone Security: Issues and Challenges", 2021.

extraordinary measures.[23] To describe ST, one must identify its constituent elements. The definition of ST is as follows:

> The discursive process through which an intersubjective understanding is constructed within a political community to treat something as an existential threat to a valued referent object and to enable a call for urgent and exceptional measures to deal with the treat.[24]

The definition explained illustrates how a particular issue transitions from being a political concern to being perceived as a threat to existence.[25] It's also important to understand that communication regarding these threats is not static. However, a dynamic process escalates the perceived threat, ultimately resulting in a tailored emergency response specific to the given situation.[26] This transformation of the threat is orchestrated by the actor who aims to present it as a matter of security, highlighting the importance and urgency of the threat to an audience, regardless of their stance on the threatened object. Bester and Arendse highlights that an existential threat can endanger the very survival of the object it targets through the basis of the security narrative that is executed.

The threat to human survival has been noted by Rogers and Kunertova, who highlight that drones are, considered to be a security threat as they have been used more frequently by non-state actors to pursue certain agendas. Rogers and Kunertova suggest that drone attacks have increased in lethality, cargo, and target range.[27] According to Majeed et al., drones pose significant security challenges to the following areas: (1) military operations and installations, (2) critical infrastructure, (3) airports, (4) police services, (5) public activities, (6) civilians and (7) prisons.[28] Therefore, the argument can be made that while the application of drones may have several benefits, it may also pose significant security challenges to the mentioned areas and actors. The authors suggest that the difficulty in identifying drones as a potential threat does not guarantee the implementation of emergency response measures. This lack of preparedness is also evident in the work of Bester and Arendse, who argue that the South African context has not recognised cyber threats as significant enough to prompt emergency responses

[23] Lucke, R.J., "How Securitization Theory Can Benefit from Psychology Findings" (Conference Presentation, ECPR General Conference, Prague, Czech Republic, 7–10 September 2016), in *Political Psychology: Methodological and Theoretical Contributions*, https://opus4.kobv.de/opus4-uni-passau/frontdoor/deliver/index/docId/621/file/Lucke_Robin_Securitization_Psycholo gy.pdf.
[24] Buzan, B. and Waever, O., *Regions and Powers: The Structure of International Security* (Cambridge University Press, 2003), 491.
[25] Bester and Arendse, "Measuring Cybersecurity Awareness: Exploring the Reliability and Factor Structure of the Cyber Security Questionnaire in a South African Military Sample", 2024.
[26] Ibid; Buzan and Waever, *Regions and Powers: The Structure of International Security*, 2003.
[27] Rogers, J. and Kunertova, D., "The Vulnerabilities of the Drone Age: Established Threats and Emerging Issues out to 2035" (Research Report, Collaboration between Centre for War Studies and ETH Zürich, 2022), https://css.ethz.ch/content/dam/ethz/special-interest/gess/cis/center-for-securities-studies/pdfs/NATO_VDA_Policy_Report.pdf.
[28] Majeed, et al., "Drone Security: Issues and Challenges", 2021.

from security-focused entities. Allworth corroborates this by stressing that UAVs and UASs are still considered an emerging phenomenon and require understanding and rigorous exploration in the SA armed forces context.[29] This does not explicitly mention the use of drones in the ST process.[30] However, this paper explores the application and dangers of drones and how they can be viewed through the same theoretical lens as a gateway to understanding how threats are conceptualised and responded to. Floyd notes that the securitisation perception of the threat is more important than the actual terminology used to describe specific emergency measures or what the threat entails.[31] Therefore, from a ST perspective the threat to the various areas pointed out by Majeed et al., may not require deeper conceptual meanings attached to their survival, instead claiming legitimacy does not depend on the audience perceptions, but rather what the securitising actor deems as legitimate.[32]

Expanding the discussion on threat legitimacy, Philipsen highlights that not all threats can be classified as existential or demand an emergency response.[33] The role of the state is considered an important facet of the securitisation process, and its role is central to its emphasis on how domestic and international issues are threatened and addressed by its existential measures.[34] Eroukhmanoff adds that ST offers a rational process through which security procedures are announced and elevated to deal with existential threats.[35] When connecting this view to UAV's/UAS's, the chapter argues that the linear or logical process of ST might not be applicable in some contexts outside of the Western world due to the importance placed on the subject of interest. The researcher puts forth that in the context of SA, the securitising actors (the state and an authorised representative) have

---

[29] Allworth, E.M., "The South African Air Force (SAAF), Unmanned Aircraft Systems and National Security: An Exploratory Study", (Unpublished Masters Dissertation, Stellenbosch University, 2018).

[30] Bester, K.J., "Exploring the Views and Perceptions of Cybersecurity among South African Military Officers", (Unpublished Doctoral Dissertation, Stellenbosch University, 2023); Bester, K.J. and Arendse, D.E., "Trojan Horse: Data Colonialism during Covid-19 Pandemic in South Africa", in *The Covid-19 Pandemic and the Politics of Life*, ed. I. Moyo and S.J. Gatsheni, (1st Edition, Routledge, India, 2023); Bester and Arendse, "Measuring Cybersecurity Awareness: Exploring the Reliability and Factor Structure of the Cyber Security Questionnaire in a South African Military Sample", 2024.

[31] Floyd, R., "Securitisation and the Function of Functional Actors", *Critical Studies on Security* 9, no. 2 (2020), 91, https://doi.org/10.1080/21624887.2020.1827590.

[32] Majeed, et al., "Drone Security: Issues and Challenges", 2021; Floyd, "Securitisation and the Function of Functional Actors", 91.

[33] Philipsen, L., "Performative Securitization: From Conditions of Success to Conditions of Possibility", *Journal of International Relations and Development* 23 (2018), 139–163, https://doi.org/10.1057/s41268-018-0130-8.

[34] Egloff, F.J. and Cavelty, M., "Attribution and Knowledge Creation Assemblages in Cybersecurity Politics", *Journal of Cybersecurity* 7, no. 1 (2021), 1–12, https://doi.org/10.1093/cybsec/tyab002.

[35] Eroukhmanoff, C, "Securitisation theory: An introduction", in International Relations Theory, ed. S. McGlinchey, R. Walters and C. Scheinpflug, (Boston: E-International, 2018).

Fig. 1 Securitisation theory process. *Source* Bester (2024)[38]

acknowledged that terrorism and nefarious actors may threaten national security interests of the state and civil society, thus ultimately challenging SA's position on establishing National Security.[36] Bester and Arendse also notes that cyber threats may also threaten SA's stance of securing cyber sovereignty.[37] Similarly to drones, one may argue that due to the mobility of UAV's/UAS's and SA's permeable border control that it is easier for criminal activity to take place thus challenging SA's position on airspace and National Security (NS).

In terms of the process of ST concerning the threat UAV's pose for the SA armed forces context, one needs to shift attention to Fig. 1.

Figure 1 demonstrates that the securitisation process occurs in a series of steps. A tenet central to the theory described in the section above is to ensure that the securitising actor makes securitising moves, through which security utterances are related to the audience. Based on Fig. 1, the process should commence with a securitising actor who often takes the role of a Minister or actor that has power to convey and acknowledge existing threats. An example of this could be linked to the Minister of Forestry, Fisheries and the Environment, Dr. Dion George, who indicated that the use of drones by citizens for fishing purposes is illegal.[39] The threat narrative was created based on the notion that drone use for fishing pressures the marine-ecosystem, thus violating sustainability and management of marine-life.[40] The securitising move in this case was initiated by the Minister and legislation was used to acknowledge the threat, but also how drones may act as both a threat and force multiplier for actors within the security cluster. The Marine Living Resources Act was used to convey the prohibition of drones in marine areas by unregistered or unlicensed drone users. Lucke makes the case that the process of securitization involves a careful consideration of the facilitating conditions that determine the success or failure of security measures. By examining these conditions, we can

[36] South Africa, South African Defence Review 2015. https://static.pmg.org.za/170512review.pdf.
[37] Bester and Arendse, "Measuring Cybersecurity Awareness: Exploring the Reliability and Factor Structure of the Cyber Security Questionnaire in a South African Military Sample", 2024.
[38] Bester K.J., "Cybersecurity Awareness from the Perspective of the South African National Defence Force Military Officer", in *The Palgrave Handbook of Language and Crisis Communication in Sub-Saharan Africa*, ed. E. Jakaza, H. Mangeya and I. Mhute (Cham: Palgrave Macmillan, 2024).
[39] South African Government., "Minister Dion George Welcomes Ruling by Supreme Court of Appeal on Drone Fishing", 18 July 2024, www.gov.za/news/media-statements/minister-dion-george-welcomes-ruling-supreme-court-appeal-drone-fishing-18.
[40] Ibid.

gain valuable insight into the speech act itself.[41] It is important to thoroughly analyse the success of securitization processes in various contexts, taking into account both internal and external conditions. This comprehensive approach allows for a more nuanced understanding of the speech act and its effectiveness in different security contexts. Buzan et al., refer to the speech act but also highlights that the conditions in which these security utterances occur could be used to observe whether the moves were unconnected or exaggerated within the political and social contexts.[42]

By connecting Fig. 1 to the chapter, the authors argue that SA had not complied with all the required elements in the ST process, since UAVs and UASs have been considered as force multipliers and a response measure. The response use of UAVs and UASs have been identified in SA literature source, where it was used to address activities pertaining to reconnaissance and intelligence gathering etc. Keeping the example of maritime threats in mind concerning drone use, it can be argued that the securitising actor identified maritime threats as posing a threat to the state and marine life. In the context of situational awareness, particularly in the armed forces, very little emphasis has been placed on the impact of UAVs on military officers, except for addressing the moral impact of the threat posed by drones.[43] Therefore, from a broader literature perspective, the vulnerability of the human factor in relation to the psychological impact on drone operators has received very little attention.

3 Technification as a "Speech Act" in the Securitisation Process

Hansen and Nissenbaum highlight the concept of technification as a unique form of linguistic expression in their work.[44] According to Hansen, the ST process should encompass three distinctive logical frameworks to effectively interconnect all participants: hyper-securitization, daily security measures, and technification.[45] Hyper-securitization is characterised by portraying hazards as imminent catastrophes.[46] This portrayal often disregards the existence of comparable historical

[41] Lucke, "How Securitization Theory Can Benefit from Psychology Findings", 7–10 September 2016.

[42] Buzan, B., Waever, O. and De Wilde, J., *Security: A New Framework for Analysis* (Lynne Rienner, 1998).

[43] Theletsane, K.L., "Drones as an Emerging Technology in the South African Military: Challenges for African Values", *Naval College Newport*, 2022, https://apps.dtic.mil/sti/citations/trecms/AD1204794.

[44] Hansen, L. and Nissenbaum, H., "Digital disaster, cyber security, and the Copenhagen School", *International Studies Quarterly* 53, no. 4 (2009): 1155–1175.

[45] Hansen L., "Theorizing the image for security studies: Visual securitization and the Muhammad cartoon crisis", *European Journal of International Relations* 17, no. 1 (2011): 51–74.

[46] Egloff and Cavelty, "Attribution and Knowledge Creation Assemblages in Cybersecurity Politics", 2021.

incidents.[47] Everyday security practices, as an operational concept, describes the methods used by security agents to pinpoint and emphasise threats, thereby making civilians aware of their vulnerabilities.[48] Thus, personal security measures are viewed as a solution to and a cause of insecurity. These measures become relevant when individuals take actions that mitigate or highlight threats. Lastly, technification frames a security issue as heavily reliant on technical expertise, effectively rendering the political discourse neutral.[49] In the ST framework, technification involves key advisors who aid the primary security agent in crafting a narrative focused on preserving vital entities. Despite their technical prowess, such advisors often hold limited sway over the decisions and actions taken in response to security challenges.

Bourdieu emphasizes that the authority of utterance lies with those authorised to speak on behalf of a specific group.[50] Dos Santos contends that the state does not act autonomously but designates a representative with suitable credentials to address security concerns.[51] Analysis of Dos Santos's perspective reveals that the state's use of a representative to address security concerns aligns with hegemonic structures, allowing influential individuals with significant social capital to shape the development and execution of security measures. Buzan et al., assert that security measures, comprising policies and actions, are primarily devised and implemented by the state.[52] However, it can be argued that technical experts may have informed the response-driven security measures. Consequently, the technification of security issues endows technical experts with epistemic authority to assess and prioritize security challenges.[53] Philipsen introduces an alternative viewpoint, suggesting that non-political actors can participate in the securitization process by introducing new security logic to establish authority.[54] Philipsen notes that these new securitising actors may offer an expanded perspective on security issues by contextualizing threat characterization.[55] One may argue that the defence industry role players may assist the securitising actor in the formation of a threat narrative, but also in the identification of security features within the speech-act.

---

[47] Hansen and Nissenbaum, "Digital disaster, cyber security, and the Copenhagen School", 2009.

[48] Egloff and Cavelty, "Attribution and Knowledge Creation Assemblages in Cybersecurity Politics", 2021.

[49] Ibid.

[50] Bourdieu, P., *Language and Symbolic Power* (Cambridge: Harvard University Press, 1994).

[51] Dos Santos, M.C., "Identity and discourse in securitisation theory", *Contexto Internacional* 40, no. 2 (2018): 229–248.

[52] Buzan, et al., Security: A New Framework for Analysis, 1998.

[53] Egloff and Cavelty, "Attribution and Knowledge Creation Assemblages in Cybersecurity Politics", 2021.

[54] Philipsen, "Performative Securitization: From Conditions of Success to Conditions of Possibility", 2018.

[55] Ibid.

4 Securitisation of Drones in South Africa

> (O)ne key exception, and that turned out to be intelligence surveillance and reconnaissance. For all of those of you in this business, you know you can never have enough. It's a constant guess. And no matter where you end up, you always need more
>
> (Bob Work, Deputy Secretary of US Defence, 2015)

The use of drones does place security in a precarious state in that some persons are assured of security, while others, those within proximity of the target are susceptible to insecurity. The nature of violence is unescapable for armed drone use and when it is particularly aimed at neutralising suspected terrorism.[56] The drone pilot is caught only between two options, to kill or not to kill. There is no middle ground that can be established since the human is removed from the engagement and only the weapon (drone) is in the battle space. Although one can argue that the drone is still operated by the human, since the human is dependent on the drone for footage, the human cannot anticipate any other nuances in the situation.[57]

Numerous nations have implemented stringent regulations about the use of drones within civilian domains.[58] These regulations stipulate that drone proprietors must operate their devices under government-issued licenses and delineate that drones are prohibited from trespassing on private property without the express consent of the owners.[59] South Africa has been identified as one of the destinations where drones are manufactured. However, the industry is small compared to the international market. Drones in SA have been considered innovative and cost-effective (Ramdass, 2024).[60] While the use of drones has been noted as having a significant impact on the battlefield, there are some challenges from a legislative perspective that prevent the drone sector from flourishing such as the lack of clear policies and guidelines that bridge the gap between civil society and the SA armed forces context.[61] In the context of creating a threat narrative, the Defence Review 2015 indicated that the SANDF has recognized the need to develop both new and existing UAV capabilities to enhance its aerospace reach.[62]

The South African Civil Aviation Authority (SACAA) oversees and controls the operations of drones within the country. According to Meyer and Khota, the initial set of regulations was introduced in May 2015 as the Eighth Amendment of the

---

[56] Johnson, B., "Coded Conflict: Algorithmic and Drone Warfare in US Security Strategy", *Journal of Military and Strategic Studies* 18, no. 4 (2018), 35–71.
[57] Ibid.
[58] Majeed, et al., "Drone Security: Issues and Challenges", 2021.
[59] Ibid.
[60] Ramdass, N, "Drone potential hindered by regulatory hurdles, lack of funding", *Engineering News*, 29 March 2024, https://www.engineeringnews.co.za/article/drone-potential-hindered-by-regulatory-hurdles-lack-of-funding-2024-03-14.
[61] Allworth, "The South African Air Force (SAAF), Unmanned Aircraft Systems and National Security: An Exploratory Study", 2018; Theletsane, "Drones as an Emerging Technology in the South African Military: Challenges for African Values", 2022.
[62] South Africa. South African Defence Review (2015).

Civil Aviation Regulations, Part 101: Remotely Piloted Aircraft Systems, under the Civil Aviation Act 13 of 2009. Updates on the regulation commenced during the 2023 period.[63] This shows the country's commitment towards an improved regulatory framework on UAVs. Despite this snail's pace since, eight years lapsed before updated regulations could take place, SA's drone regulation balances safety, privacy and innovation.[64] Shifting the discussion towards the policy development of the UAVs in the SA armed forces context, Allworth argues that regulations focusing on military UAVs are insufficient in addressing the aspects related to (1) licencing, (2) application and (3) civil-military relations.[65] Additionally, while the development of legislation is an ongoing process, the Civil Aviation Act 13 of 2009 does not effectively address the military's role and how interoperability may occur between civil society and the armed forces context. According to Theletsane there is no existing policy in the SANDF that specifies how UAVs can be integrated and utilised.[66] In terms of positioning the development of policy and legislation in line with the theoretical argument of ST, the authors argue that policy creation and promulgation are important facets in the elevation of threats and employing responses. This promulgation is important as successful securitisation does not only depend on the securitising actor mobilising emergency responses and initiating the speech-act. Instead, the display of emergency responses may also come in the form of policy and guidelines. Bester alludes to the notion that having effective legislation and frameworks in place are important to address emerging threats and identify the key role players in ensuring security.[67]

Based on the utilisation of drones in the SANDF, it is clear that one of the organisations' key role is to advance collaborative efforts with actors in the security cluster. Based on the final step of the theory, "successful securitisation", the use of drones by the SANDF is used in emergency cases, where its main activity is to assist the SAPS with the identification of damaged areas by providing aerial photographs.[68] It is evident that UAVs have an important role in disaster management, public safety, and rescue operations by whereby real-time information is

---

[63] Meyer, J. and Khota, H., "Drone Operations in South Africa: An Overview of the Legal Landscape", *Clive Dekker Hofmeyer*, 19 June 2024, www.cliffedekkerhofmeyr.com/news/publications/2024/Practice/Corporate/corporate-commercial-alert-19-June-2024-drone-operations-in-south-africa-an-overview-of-the-legal-landscape.

[64] Fraser, L., "Jail Time Warning for Drone Users in South Africa", *Business Tech*, 26 June 2024, https://businesstech.co.za/news/technology/778754/jail-time-warning-for-drone-users-in-south-africa/.

[65] Allworth, "The South African Air Force (SAAF), Unmanned Aircraft Systems and National Security: An Exploratory Study", 2018.

[66] Theletsane, "Drones as an Emerging Technology in the South African Military: Challenges for African Values", 2022.

[67] Bester, "Exploring the Views and Perceptions of Cybersecurity among South African Military Officers", 2023.

[68] Martin, G., "Drones Aid SANDF Disaster Relief Efforts in KwaZulu-Natal", *defenceWeb*, 23 May 2022, www.defenceweb.co.za/aerospace/aerospace-aerospace/drones-aid-sandf-disaster-relief-efforts-in-kwazulu-natal/.

able to accelerate search and rescue efforts.[69] In terms of research related to how UAVs and UASs are utilised in conflict areas, very little has been published on this activity. However, the historical information related to how drones have been used relate to (1) how the Seeker UAV has been used to gather intelligence during the 1987/88 operations in South eastern Angola (2) collateral missions with the SAPS focusing on crime detection.[70]

5 Drones in the SA Military Context

SA has had a long history of utilising drones.[71] The SA Air Force 10th Squadron has been identified as a key role player in the development and management of drones for mission purposes. However, the 10th squadron is nonexistent based on recent records.[72] However, despite this report on inoperability the SANDF does in fact use UAV's to not only engage in specialised military operations, but also assists the South African Police Services In multi-disciplinary relief efforts.[73]

The SANDF utilises a range of specialised military UAVs, which include the Seeker 400 and the Civet.[74] These UAVs are developed by local armed forces industry partners such as Paramount Advanced Technologies and Denel, showcasing the nation's prowess in advanced military technology.[75] The showcase of advanced military technology is generally presented at the African Aerospace and Defence Expo.[76]

According to Allworth, the SANDF uses UAS to carry out the following ten activities: (1) conducting surveillance, (2) reconnaissance, (3) conducting an assessment of damage that occurred on the battlefield, (4) military site inspections, (5) ground targeting assistance, (6) to monitor anti-aircraft artillery training whereby target drone applications are applied, (7) carrying out test and evaluation of air-launched weaponry, (8) offering support to the media such as filming

[69] Mohsan, S.A.H., Khan, M.A., Noor, F., Ullah, I. and Alsharif, M.H., "Towards the Unmanned Aerial Vehicles (UAVs), A Comprehensive Review", *Drones* 6, no. 6 (2022), 147, https://doi.org/10.3390/drones6060147.

[70] South African Air Force Website., "Air Force: The Seeker 1", www.saairforce.co.za/the-air force/aircraft/160/seeker-1 (all websites cited in this publication were last accessed and verified on 5 July 2024).

[71] Military Africa., "South Africa Military Drones—Unmanned Aerial Vehicles (UAVs)" *Military Africa*, 8 June 2022, www.military.africa/2022/06/south-africa-military-drones-unmanned-aerial-vehicles-uavs/.

[72] Ibid.

[73] Martin, "Drones Aid SANDF Disaster Relief Efforts in KwaZulu-Natal", *defenceWeb*, 23 May 2022.

[74] Allworth, "The South African Air Force (SAAF), Unmanned Aircraft Systems and National Security: An Exploratory Study", 2018.

[75] Ibid.

[76] African Aerospace Defence Expo., "African Aerospace Defence Expo", www.aadexpo.co.za/exhibitors.

the State of the Nation Address (9) conducting routine perimeter security checks, (10) border patrol and anti-poaching support through surveillance operations.[77] Adding to the abovementioned on the SANDF's capability, according to Military Africa, SA has played an important role in the military drone sector for a prolonged period.[78] There are popular defence organisations such as Milkor, Denel Dynamics, CSIR, and ATE that have played a significant role in the advancement of a variety of UAV's that include hand-launched variants to larger armed models. The SANDF actively deploys drones, such for the purpose of border patrol and peacekeeping purposes.[79] Smaller UAVs are used for surveillance purposes whereas larger drones are used for anti-poaching campaigns and for the conservation of wildlife.[80]

An aspect that is gaining momentum is that of the mental health of drone pilots, particularly those operating in the military.[81] In a recent study by Molenkamp et al., it was found that drone pilots in the military were susceptible to psychological distress.[82] There is a 'double weakness' present in the psychological distress of drone pilots operating in the military due to their distance from the actual mission and still experiencing the distress of the actual mission. Through this they encounter stigma as they are not regarded as actual pilots in that their proximity from the mission undermines their contribution according to traditional military culture that sees aircraft pilots as 'real pilots'.[83] Although South Africa is not yet at this juncture, it stresses the need for military psychologists to become engaged in drone missions. Since there is an element of mental health that will leave the drone pilot vulnerable and compromise his ability to perform his task, the military psychologist serves to keep him psychologically fit for the mission, albeit a drone mission. The act of war and killing may have shifted in terms of proximity, but this does not safeguard soldiers from becoming prone to different forms of Post-Traumatic Stress Disorder (PTSD). As such, in thinking through operationalizing

---

[77] Allworth, "The South African Air Force (SAAF), Unmanned Aircraft Systems and National Security: An Exploratory Study", 2018.

[78] Military Africa, "South Africa Military Drones—Unmanned Aerial Vehicles (UAVs)" *Military Africa*, 8 June 2022, www.military.africa/2022/06/south-africa-military-drones-unmanned-aerial-vehicles-uavs/.

[79] Ibid.

[80] Ibid.

[81] Holz, J., "Victimhood and Trauma within Drone Warfare", *Critical Military Studies* 9, no. 2 (2021), 175–190, https://doi.org/10.1080/23337486.2021.1953738; Molenkamp, et al., "A Playstation Mentality to Killing? Adverse Psychological Consequences in Drone Pilots and the Stigmatization Thereof in the Military", 2024.

[82] Molenkamp, et al., "A Playstation Mentality to Killing? Adverse Psychological Consequences in Drone Pilots and the Stigmatization Thereof in the Military", 2024.

[83] Holz, J., "Victimhood and Trauma within Drone Warfare", *Critical Military Studies* 9, no. 2 (2021), 175–190, HYPERLINK "sps:urlprefix::https" https://doi.org/10.1080/23337486.2021.1953738; Molenkamp, et al., "A Playstation Mentality to Killing? Adverse Psychological Consequences in Drone Pilots and the Stigmatization Thereof in the Military", 2024.

of drones within the South African military, there should be a serious consideration of the military psychologist as part of the mission readiness.

6 Recommendations

The social implications of UAV research in the SA armed forces context have not yet been fully established. Therefore, despite the Minister of Defence highlighting the development of UAVs in the SANDF, the implications of its use and its utilisation as a force multiplier must be explored further. This chapter aimed to provide a contextual and theoretical lens through which UAVs/UASs can be viewed and in doings so contributed to the existing body of literature on UAVs and UASs within the SA context.

In terms of viewing the available developments and literature on UAVs and UASs in the SA armed forces context. Future research should explore how the budgetary implications may have an impact on the implementation and use of unmanned aircraft and systems. These technologies assist in the advancement of the SANDF capabilities within the aerospace domain, but also from a national security perspective. Furthermore, the chapter recommends from a practical perspective that future research may focus on the various size and capabilities of UAV and UASs in relation to its cost. More specifically, it should be highlighted that research be geared towards understanding the feasibility of UAVs in the SA armed forces context given the continued budgetary constraints. This recommendation also aligns with Allworth on exploring the feasibility of unmanned vehicles in a developing armed forces context.[84]

Moreover, the ST framework employed in this chapter could only be assessed from a process perspective, therefore not all facets of the theory could be unpacked such as the philosophical underpinnings of the speech act theory and how it facilitates the threat narrative. Thus, future research may need to investigate the trends in which drones as a security measure may expand on the understanding of how threats are elevated, and the unique response employed by nation states and the armed forces. In addition, the chapter showed that the securitising actor is influenced by technical experts within the aerospace industry. However, the threat narrative issued and how the use of UAVs and UASs are framed have not been clearly conveyed by securitising actors within the SA armed force context. Thus, additional exploration may need to shift to how conceptualisations attributed to emerging technology such as "drones" are understood by military members and how these unmanned crafts and systems are able to advance the SANDF to engage in future forms of warfare.

[84] Allworth, "The South African Air Force (SAAF), Unmanned Aircraft Systems and National Security: An Exploratory Study", 2018.

The work of the military psychologist may also be shifting in the advancement of technology and the inclusion of drones. For this reason, the military psychologist within the South African military may need to become familiar with the work of the drone pilot and the impact as well as stressors the drone pilot may experience. Each context is unique and in South Africa, drone pilots may experience psychological distress in a different manner or due to diverse circumstances. It is therefore vital to engage in research and practice that is context-specific and that caters for the South African soldier.

7 Conclusion

This chapter concludes that drones in the SA armed forces context significantly contribute to providing the military with real-time information, surveillance, and reconnaissance in hostile territories. It is also apparent that there is a security concern related to the civil-military relationship concerning the use of drones. The positionality of SA is that legislation is implemented to regulate the use of drones used for commercial and private use. However, in terms of legislation on drone use in the armed forces context, there remains very little published in the SA armed forces context. This gap legislation gap does not bode well for the theoretical argument of ST. UAVs in the SA armed forces context in terms of legislation needs to be more synchronised with those guidelines and regulations for civil society. This synchronicity is to ensure that cross collaboration occurs within the SA security cluster and the effective regulation of UAVs/UASs in the SA context. Furthermore, the intensified budgetary constraints imposed on the SANDF makes the acquisition of newer drones challenging. The budgetary constraint may ultimately result in a diminished organisation that is unable to have effective domain awareness systems that are able to cope with the demands of operations. Moreover, from a National Security point of view, budgetary constraints not only prevent effective advancement in the armed forces context, but also adds to the crippling stance that the organisation may not be able to prevent future attacks from more complex UASs.

This chapter presented that not all the boxes have been checked according to the process model presented by the author. Globally there has been an acknowledgement on threats that pose significant challenges to the safety of the military member and strategic operations. Within the SA armed forces context, there has been limited communication on the threats to its members and operations that would necessitate the use of drones. While the use of drones may not be indicative of an emergency response, it does however link with the notion that security measures does not require an emergency response, a declaration by the securitising actor and the execution of a speech act.

Acknowledgements This work is based on research supported by the National Institute for the Humanities and Social Sciences (NIHSS).

Dr. Kyle John Bester is a registered Research Psychologist and senior psychology lecturer at the University of South Africa. He holds a master's degree in research psychology from the University of the Western Cape. He supervises postgraduate students who have taken up research related to cybersecurity, online security behaviour, artificial intelligence and digital culture. He specialises in research focused on cybersecurity awareness in the South African armed forces context. He has a PhD in Military Science from Stellenbosch University. He serves as an executive committee member in the Division for Research and Methodology (DRM) at the Psychological Society of South Africa (PsySSA). Dr Bester is an emerging researcher, and his research interests include military science; data-colonialism; cybersecurity awareness; securitisation of cyberspace and online security behaviour.

Danille Elize Arendse obtained her PhD in Psychology from the University of Pretoria and also holds a BA Honours in Psychology and an MA in Research Psychology from the University of the Western Cape. She became a uniformed member of the South African National Defence Force (SANDF) in 2011, where she worked as a Research Psychologist at the Military Psychological Institute (MPI). Now a Major, Arendse has previously taken on the role of Research Psychology Intern Supervisor and Coordinator at MPI. Additionally, she serves as a Research Associate in the Department of Psychology at the University of Pretoria and is an Accredited Conflict Dynamics Mediator. In 2022, Arendse received the Diverse Black Africa Research and Travel Grants from Michigan State University, and in 2023, she was awarded the prestigious NIHSS/SU postdoctoral fellowship at the Centre for the Study of the Afterlife of Violence and the Reparative Quest (AVReQ) at Stellenbosch University. Arendse has presented and published papers both locally and internationally and currently practices as a Research Psychologist at MPI.

An Eye Above the Water's Surface: Drones in the African Maritime Domain

Tshegofatso Johanna Ramachela

Abstract

Drones have the potential to serve as force multipliers, to supplement the role of traditional navies and coastguards by operating in tandem with these forces. In this sense, unmanned systems would not replace traditional maritime forces, but rather they would play a complementary role. The necessity to integrate this technology in Africa cannot be understated, especially given the lacking capacity of several African navies. Moreover, this technology is increasingly being used by nefarious actors. This chapter explores how certain African stakeholders, both state and non-state have developed and/or deployed drones in the maritime domain for various uses. The main countries discussed include Seychelles and Nigeria, as well as an assessment of the drone deployment in the Mediterranean Sea. Leading from this, the focus is on the ways in which non-state actors have deployed drones in the maritime environment. To sum up the chapter, the discussion revolves around the future trends, opportunities and implications of drone development and deployment in Africa's maritime domain.

1 Introduction

Drones in the maritime domain are not a new development; however, they are certainly a new frontier for African countries. As more states recognise the advantages of drones, there has been more interest from African countries, particularly African navies and coastguards, to acquire and deploy unmanned aerial vehicles (loosely labelled as drones) within their territorial waters. The maritime domain

T. J. Ramachela (✉)
Ocean Regions Programme, University of Pretoria, Pretoria, South Africa
e-mail: tshego.ramachela@gmail.com

W. S. Coetzee and D. Putter (eds.), *Drones in the African Battlespaces*, Southern Space Studies, https://doi.org/10.1007/978-3-031-87147-4_7

has long been considered a strategic space and is increasingly recognised as a vital part of international security agendas. The maritime domain has no shortage of security challenges, with threats including but not limited to piracy which occurs on the high seas and armed robbery which occurs in territorial waters of littoral states. Securing maritime spaces, including trade and shipping routes, otherwise known as sea lines of communication (SLOCs), the marine environment, seafarers, and shoring up against landward-bound insecurity is the prerogative of several stakeholders. While each state is responsible for securing its own territorial waters, the areas outside this, known as the high seas or international waters, constitute a shared space where no one state has jurisdiction. Thus, within this space, various actors, state, inter-state, and non-state actors, such as private security entities, take it upon themselves to preserve their interests and security. In addition to traditional navies securing geopolitical interests and fulfilling maritime diplomatic duties, constabulary forces such as coastguards that patrol and protect sea lanes as well as maritime boundaries are part and parcel of maritime security.

This is a major challenge for African littoral and coastal countries, most of which lack sufficient capacity to effectively manage and enforce maritime safety and security, for several reasons. First, for a long time, Africa has suffered from what is known as sea blindness, a failure or rather overemphasis and prioritisation of land-based security issues at the expense of littoral and/or offshore insecurity. This is gradually changing, as this chapter will detail. Second, the existing maritime capabilities of some African states are disproportionate to the continent's extensive maritime domain. Thirty-nine of the fifty-five countries in Africa are coastal; the continent has a combined maritime coastal length of 26,000 nautical miles, with a combined exclusive economic zone (EEZ) of 13 million square kilometres.[1] The Atlantic Ocean lies to the West of Africa, while the Indian Ocean borders Africa's eastern coastline. South of Africa is the Antarctic or Southern Ocean, in the north lies the Mediterranean Sea, and to the north-east is the Red Sea. In addition to an expansive network of inland bodies of water, namely lakes and rivers, many strategic trade (shipping) routes and chokepoints exist in Africa's combined maritime space. These include the Suez Canal, the Bab-El-Mandeb Straits, the Cape of Good Hope, the Mozambique Channel, and the Straits of Gibraltar.[2] The bulk of the world's trade crosses these waterways daily.[3] Thus, they are of significant importance not only for Africa but extra-regional (external non-African) actors as well. Within this vast space, there are a plethora of maritime security issues, also referred to as maritime crime: maritime terrorism, piracy, armed robbery, kidnapping for ransom of seafarers, transnational organised crime,

[1] Surbun, V., "Africa's combined exclusive maritime zone concept", *Institute for Security Studies*, 15 March 2021, https://issafrica.org/research/africa-report/africas-combined-exclusive-maritime-zone-concept.

[2] Lauriat, G., "Global maritime choke points", *American Journal of Transportation* 724, 8 (19 April 2021).

[3] Nehra, A., "Strategic significance of Mozambique channel", *CESCUBE*, 29 May 2021, https://www.cescube.com/vp-strategic-significance-of-mozambique-channel.

trafficking of narcotics, human trafficking, and trafficking of illicit goods, such as arms.[4] Other threats include illegal, unreported and unregulated (IUU) fishing and other environmental crimes, such as waste dumping and maritime accidents, namely ship grounding, oil spills and disasters at sea.[5] All these activities have an expansive impact on local and regional economies and, thus, on the geopolitical stability of the continent.

With such cross-cutting challenges, a vast expanse of territory and limited air and naval capabilities and resources to enforce maritime security safeguards, many African countries are exploring the potential utility of drones to bridge this gap. On land, drones in Africa have been instrumental in carrying out significant activities, such as anti-poaching operations, search and rescue missions, and delivering life-saving medical materials and humanitarian supplies.[6] Thus, it stands to reason that at sea, drones can serve as force multipliers, to supplement inefficiencies and complement the role of traditional naval and maritime forces. Using empirical cases of Seychelles, Nigeria, two countries leading maritime drone applications in Africa, and drone deployment in the Mediterranean Sea, this chapter unpacks the trend of maritime drone proliferation in Africa. In doing so, the chapter assesses the potential benefits and risks and subsequently concludes with a critical analysis of the future opportunities, challenges, and implications of maritime drone deployment.

2 What Distinguishes Maritime Drones from Typical Drones?

Unmanned aerial vehicles (UAVs), commonly known as drones, differ in structure (design), size and functions. The weight and size of a drone affect its operational utility. That is, larger, heavier drones consume more energy per unit of time and/or distance and cannot fly as efficiently or as far and long as smaller, lighter models.[7] As distinct from manned aircraft, drones do not require a human operator, rather, they are autonomous and/or remotely controlled. They can either be based onshore or on a maritime platform.[8] In the latter case, the vessel may serve as a command

[4] Amani Africa, "The state of Maritime Security in Africa", Insights on the Peace and Security Council (PSC), 23 July 2021, https://amaniafrica-et.org/the-state-of-maritime-security-in-africa/.
[5] Marie, A. and Bueger, C., "Seychelles: Island Solutions and Capacity Building Successes", in *Capacity Building for Maritime Security: The Western Indian Ocean Experience*, ed. Bueger C., Edmunds T. and McCabe R. (Cham: Springer International Publishing, 2021), 199–222, https://doi.org/10.1007/978-3-030-50064-1_8.
[6] Child, D., "From Kigali to Khartoum: Africa's drone revolution", *Aljazeera*, 30 January 2018; Rawley, C. and Patmon, C., "Drones in Africa: a leap ahead for maritime security", *Center for International Maritime Security*, 9 October 2018, https://cimsec.org/drones-in-africa-a-leap-ahead-for-maritime-security/.
[7] Amprius, "What affects the range of a drone?", 10 August 2023, https://amprius.com/drone-range/.
[8] Rawley and Patmon, "Drones in Africa: a leap ahead for maritime security", 9 October 2018.

centre. Various other methods of control and deployment are still being tested. In 2022, for example, Turkey launched its first-ever Landing Helicopter Dock vessel (LHD)—the TCG Anadolu, which will also carry and deploy drones.[9] Ankara plans to develop more naval carrier platforms for export purposes,[10] and Africa is an ideal market. In the maritime domain, semi-autonomous drones, sometimes referred to as maritime autonomous vehicles (MAVs), are designed to withstand the oceanic environment. The current systems, both fixed and rotary winged types, can be adapted for the purpose of surveillance in the maritime domain with minimal costs incurred.[11] These drones can be deployed for a range of maritime security functions, namely surveillance, scanning remote or inaccessible areas, combat piracy and maritime terrorism, oceanographic mapping and marine fauna surveys, monitoring ports, offshore oil and gas facilities, coastal security, search and rescue, border policing and even the inspection of ships and cargo.[12]

Drones are considered low risk and high reward; they are more accessible and offer a relatively low-cost alternative to conducting maritime domain surveillance. Unlike other aerial assets, maritime drones fly at a lower altitude close to the water's surface and within proximity to coastal facilities, and thus, collisions with civil aviation aircraft are largely avoidable.[13] Because they operate in open-air assemblies over vessels belonging to various entities, consistent monitoring of flightpaths vis-à-vis other aircraft and persons on the ground is necessary.[14] Their relative ease of deployment and the low cost of acquisition in comparison to traditional aircraft make drones especially appealing to African security forces with limited funding and capabilities. Moreover, deploying helicopters in harsh coastal weather conditions can be dangerous and risk the lives of those onboard.[15] The ability of drones to operate remotely, in contested areas, and provide surveillance

[9] Brimelow, B., "Shunned by the US, Turkey is preparing to launch its first aircraft carrier—but it will come with a twist", Business Insider, 2023, https://www.businessinsider.com/turkey-preparing-for-first-aircraft-carrier-tcg-anadolu-with-drones-2023-2.

[10] Iddon, P., "Turkey has a new drone to use with its new drone aircraft carrier, and it's looking to sell it to the rest of the world", Business Insider, 28 December 2023, https://www.businessinsider.com/turkey-wants-to-export-drones-and-drone-carrying-ship-2023-12.

[11] Rawley and Patmon, "Drones in Africa: a leap ahead for maritime security", 9 October 2018.

[12] Dyer, J., Defending African Maritime Sovereignty with Finite Resources: The Prospects and Power of Drones and AUVs (Cape Town: Blue Economy Future, 2018), https://www.blueeconomyfuture.org.za/gallery/African%20maritime%20drones%20Jan%202021.pdf; Allen, K., "Drones and Violent Nonstate Actors in Africa", Africa Center for Strategic Studies, 7 November 2022, https://africacenter.org/spotlight/drones-and-violent-nonstate-actors-in-africa/.

[13] Rawley and Patmon, "Drones in Africa: a leap ahead for maritime security", 9 October 2018.

[14] Smith-Godfrey, S., "The use of drones in ports", presentation at Virtual Southern African Transport Conference, Pretoria, 5–7 July 2021.

[15] African Defence Forum, "Eyes in the Sky", *ADF Magazine*, 2 February 2023, https://adf-magazine.com/2023/01/eyes-in-the-sky/.

and real-time intelligence whilst minimising the risk to human lives has dramatically transformed naval operations as we know them.[16] Given the wide-ranging benefits, drones are the logical next step in Africa's technological maritime security evolution.

3 African Maritime Security and the Growing Interest in Drones

Historically, the maritime sector in Africa remains underdeveloped in terms of offshore (maritime) capabilities, facilities, learning and employment opportunities, rendering Africa vulnerable to what is referred to as "sea blindness". However, this is changing as more countries recognise the importance of the maritime domain and the necessity to secure not only their territorial waters but their marine resources as well.

Millions of Africans are dependent on marine resources and maritime activities, such as fishing, as a source of commerce and consumption.[17] Along with fisheries, coastal tourism, maritime transport, and renewable ocean energy are part and parcel of the blue economy. In 2018, Africa's blue economy generated approximately $300 billion USD and approximately 49 million jobs.[18] This figure is projected to increase, granted that enhanced ocean governance is prioritised and effectively administered. Africa's blue economy has the potential to uplift millions out of poverty and drive the economic and sustainable development of the continent. Unfortunately, maritime insecurity presents a major challenge to the growth of Africa's blue economies and, in turn, the growth and prosperity of the continent. Therefore, with this untapped potential and the challenges hindering the development of Africa's blue economy, safeguarding the maritime domain is a strategic priority for many African countries.

As previously mentioned, most African coastal and island countries have limited offshore patrol capabilities, and some lack sufficient aerial assets.[19] That being the case, drones have been identified as a potential solution to this predicament. Africa's private sector, among them tech start-ups and private security firms, was

---

[16] Prasanga, A.P.A., "The strategic impact of military drone proliferation on Indo-Pacific maritime security", *Center for International Maritime Security*, 23 November 2023.

[17] African Natural Resources Centre, The future of marine fisheries in the African Blue Economy (Abidjan: African Development Bank Group, 2022), https://www.afdb.org/sites/default/files/documents/publications/afdb_the_future_of_marine_fisheries_in_the_african_blue_economy_-_final_for_publication.pdf.

[18] World Bank Group, Blue Economy in Africa: A Synthesis (Washington DC: World Bank Group, 2022), https://thedocs.worldbank.org/en/doc/213a25f8770328e39b2ef15e7104a136-0320012022/original/Overview.pdf.

[19] Reva, D. and Ramachela, T.J., "Can a new wave of drone tech make Africa's seas safer?", *Institute for Security Studies*, 30 November 2022, https://issafrica.org/iss-today/can-a-new-wave-of-drone-tech-make-africas-seas-safer.

the first to use drone technology in the maritime domain for surveillance, monitoring ports, offshore oil and gas facilities, oceanographic surveys and detecting marine pollution.[20] Therefore, it is possible for state actors, namely maritime security authorities, to explore the benefits.

Obtaining drone technology and deploying it at sea could be quite complex due to technological challenges, financial constraints, legal regulations, or lack thereof. A possible solution[21] offers is outsourcing to Western partners in the form of "contractor-operated drones". There are also viable non-Western partners with the expertise to manufacture and supply drones and provide technical assistance. Many of these stakeholders already supply a number of African countries with drones,[22] Turkey, for example, has supplied several of its Bayraktar TB2 drones, ANKA-S and Aksungur drones to Togo, Egypt, Morocco and Tunisia. In the long term, however, the practice of outsourcing drone operations is unsustainable and detrimental to autonomous decision-making and policy formulation. To bridge the prevailing security gap, some African maritime authorities, primarily West African navies and coast guards of island states, have included plans to bolster their capacity through the acquisition of additional vessels and more modern systems.

It is worth noting that the literature and reporting on drone procurement and maritime use in Africa are limited, perhaps due to the nascence of the field. However, as the demand for maritime drones increases, so does academic scholarship on drones for maritime safety and security in Africa. Some authors[23] are delving deeper into the subject of drones for maritime security, and they are not alone in their academic pursuits. Allworth[24] argues that unmanned autonomous systems (UAS) could serve as force multipliers in maritime surveillance operations of the SANDF.

Drones are becoming part and parcel of South Africa's maritime security strategy.[25] Meredith[26] proposed using unmanned aircraft systems, i.e. drones, to carry out South Africa's Long-Range and Long-Endurance (LRLE) maritime search and rescue (SAR) operations. In the same vein, Mota[27] argues that the use of drones for

[20] Ibid.

[21] Rawley and Patmon, "Drones in Africa: a leap ahead for maritime security", 9 October 2018.

[22] Crisis Group, "Turkiye's growing drone exports", 20 December 2023, https://www.crisisgroup.org/europe-central-asia/western-europemediterranean/turkiye/turkiyes-growing-drone-exports.

[23] Klein, N., "Maritime autonomous vehicles and international laws on boat migration: lessons from the use of drones in the Mediterranean", Marine Policy 127, no. 1 (2021): 1–10, https://doi.org/10.1016/j.marpol.2021.104447.

[24] Allworth, E.M., "The South African Air Force (SAAF), unmanned aircraft systems and national security: an exploratory study", Stellenbosch University, 2021, http://hdl.handle.net/10019.1/109960.

[25] Dyer, *Defending African Maritime Sovereignty with Finite Resources*, 22.

[26] Meredith, A.P., "An unmanned aircraft system for search and rescue", Stellenbosch University, 2011, http://hdl.handle.net/10019.1/6720.

[27] Mota, A., "Enhancement of search and rescue missions in the West coast of Africa: examining the possibilities of use of drones in Cabo Verde search and rescue", World Maritime University, 2023.

SAR operations in Cabo Verde is crucial to facilitate and enhance the efficiency of maritime SAR operations. Despite the challenges borne of financial implications, Mota[28] points out that drones for maritime SAR can improve the chances of survival and locate both people and goods in expansive territory. However, while drone SAR operations have proven successful on land, this approach at sea still requires extensive testing because whilst a drone can locate humans and/or objects, it alone cannot carry out the rescue effort (yet). Even with their wide-ranging potential, drones cannot currently replace traditional maritime vessels.

There are other areas being explored, such as using drones to survey coastal safety. Hlabela et al.[29] note that the use of drones to photograph and assess breakwaters (coastal barriers to protect a harbour or beach) has become a topical theme in the research community. Undoubtedly, both practically and academically, the interest in maritime drones and their potential is growing.

While maritime drones have yet to be fully adopted, operationalised and regulated in Africa, they have, to some extent, been integrated into maritime safety and security activities. To date, drones have primarily been deployed as a means to collect data, particularly to combat maritime crime.[30] A number of analysts and practitioners believe that drones have the potential to deter and/or decrease illicit activity at sea. This may be true, but there are other factors to consider. African countries that have the ambition to deploy drones in their maritime domains must take into consideration the cost of acquisition and maintenance as well as the infrastructure required to effectively use drones for maritime security.[31] Smith-Godfrey,[32] for example, points out that in order for states to employ maritime drones to deter environmental crime such as illicit waste dumping or any other crime that may adversely harm marine ecosystems, they must develop a framework for implementation.

The first step of drone operation in the maritime domain is to license the drone, including its "flight paths and patterns approvals".[33] The latter is especially important where ports of entry are concerned. In ports, drones are typically deployed under the supervision of a port engineer(s) who then oversees the drone operations.[34] In addition to the port engineers, drone operators must be trained and equipped with the relevant skills. According to Hlabela et al.,[35] port operations and productivity can be surveyed using data collected by drones in tandem with the collection of environmental data. This will obviously require each drone to be equipped with the correct sensor suite. The use of drones in this regard would

---

[28] Ibid., 39.

[29] Hlabela, P., Harribhai, J. and Seboqa, M., CSIR Final Report: Smart Ports (Pretoria: CSIR, 2016), http://hdl.handle.net/10204/11472.

[30] Smith-Godfrey, *The use of drones in ports*, 1.

[31] Rawley and Patmon, "Drones in Africa: a leap ahead for maritime security", 9 October 2018, 1.

[32] Smith-Godfrey, *The use of drones in ports*, 3.

[33] Ibid.

[34] Hlabela et al., *CSIR Final Report: Smart Ports*, 12.

[35] Ibid., 4.

decrease delays in inspections whilst minimising the danger to employees. Unfortunately, many African ports do not have modern enough infrastructure to support and manage shipping traffic, let alone the integration of drone technology. Thus, it is not only a matter of acquiring drone systems but also upgrading older systems and modernising outdated infrastructure. The adaptation of ports into smart ports (including the use of artificial intelligence) would be a major game changer for Africa vis-à-vis maritime safety, security, sustainable development and resilience. Acquiring drone technology is the crucial next phase of the process.

4 Manufacturers of Maritime Drones

Having recognised the potential of drones to enhance maritime security, African countries have increasingly sought to procure this technology. As a result, the demand for drones with maritime applications has risen significantly, as African countries attempt to enhance their defence capabilities and acquire a competitive edge. This has made Africa a highly profitable market, which attracts partnerships and investments from various stakeholders, both state and non-state, with the capacity to manufacture and supply drones. Consequently, Africa's drone market is now an amalgamation of an assortment of technology supplied by multiple countries.[36] The main exporters (suppliers) of drones to Africa are the United States (US), the United Kingdom (UK), China, Turkey, the United Arab Emirates (UAE), Israel, South Korea, Iran and South Africa.[37] Turkey, in particular, is developing into a major defence partner in Africa. Turkey is gradually bolstering its defence diplomacy efforts and continues to supply advanced weaponry, including drones, to African countries. Turkey's Bayraktar TB3 naval drones are currently in high demand due to the explosive media coverage it received in the first year of the Ukraine-Russia war.[38]

Several African and Africa-based companies manufacture drones for maritime purposes, including but not limited to the South African Denel, Milkor and the Moroccan technology start-up Atlan Space.[39] Atlan Space provides ocean monitoring services, including satellite imaging, tracking data from vessel monitoring systems, in situ measurements, remote sensing, and data from automatic identification systems (AIS). Among their numerous services and industries, Atlan Space's

[36] African Defence Forum, "The dizzying possibilities of drones in Africa", *defenceWeb*, 20 October 2023, https://www.defenceweb.co.za/aerospace/aerospace-aerospace/the-dizzying-possibilities-of-drones-in-africa/.

[37] Hochet-Bodin, N., "Africa, the new playground for drone exporters", *Le Monde*, 5 January 2024, https://www.lemonde.fr/en/international/article/2024/01/05/africa-the-new-playground-for-drone-exporters_6405337_4.html; Lake, J., "Africa's UAV defence imports", Times Aerospace, 2022, https://www.timesaerospace.aero/features/defence/africas-uav-defence-imports.

[38] Iddon, "Turkey has a new drone to use with its new drone aircraft carrier", 28 December 2023.

[39] Military Africa, "South Africa military drones—Unmanned Aerial Vehicles (UAVs)", 8 June 2022, https://www.military.africa/2022/06/south-africa-military-drones-unmanned-aerial-vehicles-uavs/.

ocean monitoring industry developed smart drones to detect and deter maritime-related crimes, including but not limited to, illegal fishing and oil spills in the territorial waters of Morocco and Africa as a whole.[40] Atlan Space is perhaps the leading manufacturer of autonomous platforms in North Africa. The list of manufacturers and suppliers will no doubt expand with increases in demand. Only a handful of African countries have begun using drones within their respective maritime zones. Among the top users of drones for maritime security are Nigeria and Seychelles.

5 Seychelles

In 2009, the U.S. military deployed MQ 9 Reaper drones from its base in Seychelles to monitor the Indian Ocean (a piracy high-risk area at the time) and collect photographic evidence of acts of piracy.[41] Although this drone crashed,[42] the initiative by the U.S. Africa Command (AFRICOM) demonstrates that drones purposefully used in support of maritime security are not a new development in the Western Indian Ocean. Prasanga[43] points out that the Indo-Pacific region (which includes a section of the eastern Indian Ocean) is increasingly witnessing the rapid proliferation of military-grade drones; this is now a significant dimension of contemporary maritime security dynamics.

Seychelles is among the few African countries that have and continue to deploy drones in the maritime domain. Seychelles is an island nation, an archipelago comprised of 115 islands, in the Western Indian Ocean. As an island nation with an exclusive economic zone of 137 million square kilometres, the maritime domain is inseparable from Seychelles' security and development. The Seychellois population of less than 100,000 is highly dependent on the marine environment and the resources within as a source of livelihood.[44] Maritime security and the development of Seychelles' blue economy are of notable significance. The small island nation's economy is inextricably linked to crucial maritime sectors, namely fishing, a key source of nutrition and tourism, accounting for 83 percent of its gross domestic product (GDP).[45] Thus, Seychelles has enacted protective measures to bolster maritime security for its waters. The strategic importance of its maritime

[40] Cutullic, S., "The drone industry in Africa", European Cluster Collaboration Platform, 28 March 2022, https://www.clustercollaboration.eu/content/drone-industry-africa.

[41] Onuoha, F.C., "Piracy and Maritime Security off the Horn of Africa: Connections, Causes, and Concerns", African Security 3, no. 4 (2010): 191–215, https://doi.org/10.1080/19392206.2010.533067.

[42] Hashim, A. and Patte, G., "What is that Buzz? The rise of Drone Warfare", Counter Terrorist Trends and Analyses 4, no. 9 (2012): 8–13.

[43] Prasanga, "The strategic impact of military drone proliferation on Indo-Pacific maritime security", 23 November 2023.

[44] Philpott, D., Gray, T.S. and Stead, S.M., "Seychelles, a Vulnerable or Resilient SIDS? A Local Perspective", Island Studies Journal 10, no. 1 (2015): 31–48, https://doi.org/10.24043/isj.319.

[45] Marie and Bueger, "Seychelles: Island Solutions and Capacity Building Successes", 2021.

domain and the threats plaguing it prompted the government to designate fourteen marine protected areas within its territory.[46]

Seychelles dedicates substantial resources to safeguarding its maritime environment. For years, Seychelles has been the victim of rampant illegal, unreported and unregulated (IUU) fishing.[47] In addition to IUU fishing, piracy, as well as the trafficking of illicit narcotics (such as heroin), human trafficking and arms smuggling, are among the maritime security challenges the island nation grapples with.[48] As this is the case with many African island nations, Seychelles' EEZ of roughly 1.3 million square kilometres is much larger than its land area of 455 square kilometres.[49] Despite its large EEZ, Seychelles does not have a naval capability in the traditional sense. As one of the world's smallest militaries, the Seychelles Defence Forces' primary responsibility is maritime security, namely anti-piracy operations, countering piracy and narcotics smuggling[50]—which basically equates to maritime law enforcement. Seychelles maintains bilateral maritime (defence) agreements with several countries to conduct its vital maritime security operations. In 2021, the Seychelles National Assembly approved and ratified an agreement to counter illicit transnational maritime activity with the assistance of the U.S.[51] The bulk of Seychelles' maritime security enforcement is conducted by its coast guard. Its Coast Guard has several patrol boats, supported by a limited number of aircraft with the capability to conduct reconnaissance and surveillance missions.[52]

Given its vast EEZ, limited naval capability and material capacity and finite financial and human resources, Seychelles has opted to incorporate drones in tandem with its existing maritime security mechanisms. In June 2018, in cooperation with the Moroccan drone company Atlan Space, Seychelles announced a pilot project to combat illegal fishing.[53] This initiative entailed the procurement of drones with integrated artificial intelligence (AI) to survey vast areas for environmental crimes, such as deforestation, poaching and illegal fishing. The use of

---

[46] Ibid.

[47] Payet, R., "Illegal fishing in Seychelles: a review of its implications for a small island developing state", presentation at Forging Unity: Coastal communities and the Indian Ocean's future, Chennai, India, 9–13 October 2001.

[48] Marie and Bueger, "Seychelles: Island Solutions and Capacity Building Successes", 2021, 204.

[49] Lablache, J. and Uranie, S, "Seychelles taking steps to protect more marine areas of its oceanic zone", *Seychelles News Agency*, 11 June 2014, http://www.seychellesnewsagency.com/articles/702/Seychelles+taking+steps+to+protect+more+marine+areas+of+its+oceanic+zone.

[50] Bueger, C., Adeline, R. and Cannon, B.J., "Anti-Piracy Lessons from the Seychelles", War on the Rocks, 23 April 2024, https://warontherocks.com/2024/04/anti-piracy-lessons-from-the-seychelles/.

[51] Bonnelame, B., "Maritime security expected to increase in Seychelles' waters through agreement with US", 28 May 2021, http://www.seychellesnewsagency.com/articles/14904/Maritime+security+expected+to+increase+in+Seychelles+waters+through+agreement+with+US.

[52] ADF, "Eyes in the Sky", *ADF Magazine*, 2 February 2023, https://adf-magazine.com/2023/01/eyes-in-the-sky/.

[53] Scott, K., "Drones driven by AI will track illegal fishing in African waters", *CNN*, 15 August 2018, https://edition.cnn.com/2018/08/15/africa/atlan-space-ai-drone-morocco/index.html.

AI is intended to minimise not only the cost involved in operations but also the human resources required.[54] Essentially, AI is able to substitute human operators and carry out the tasks of the pilot, data analyst(s), and transmission equipment,[55] vastly improving the cost components and reducing errors due to fatigue. It also substitutes the Coast Guard elements by conducting their monitoring tasks, thereby providing the Coast Guard with a more focused operational intent to intercept target vessels.[56] Besides the drones supplied by Atlan Space, in 2021, Seychelles' fishing authority acquired two long-range drones equipped with AI.[57] The drones were procured through the FishGuard initiative, once again in cooperation with Atlan Space, a non-profit environmental communications centre, GRID-Arendal and Trygg Mat Tracking, the Norwegian intelligence analysis firm specialising in fisheries.[58] In tandem with procuring these drones, the Seychelles Air Force pilots received the necessary training to operate the drones,[59] which is an essential part of capability integration.

According to the chief executive officer (CEO) and co-founder of Atlan Space, Badr Idrissi, the long-range drones are pre-programmed with intelligence about fishing hotspots.[60] The drones facilitate real-time surveillance of "target vessels", once the drone detects a vessel, the AI identifies the type of vessel, be it a tanker or fishing trawler.[61] In the case of the latter, the drone determines whether the fishing vessel is authorised and whether it is operating in an MPA.[62] Should the vessel be engaged in illicit activity, the drone registers the vessel's location and identification number, as well as the number of individuals on board (if possible), and transmits this information to the relevant authorities through a satellite link.[63] In addition to this, the drones monitor remote areas along the coastline making policing, search and rescue operations more effective. The extent to which these drones have been successful in real-life situations is not reported in the public domain. It remains difficult to ascertain the progress of the said operation as well as determine the type of drones used. Regrettably, this does not provide an opportunity to conduct the in-depth analysis required to enhance policy and provide strategic recommendations.

---

[54] Ibid.

[55] Ibid.

[56] Ibid.

[57] African Defence Forum, "Drones offer extra eyes in illegal fishing battle", *ADF Magazine*, 8 December 2021, https://adf-magazine.com/2021/12/drones-offer-extra-eyes-in-illegal-fishing-battle/.

[58] Ibid.

[59] Stop Illegal Fishing, "Seychelles testing drones to monitor waters for illegal fishing, potentially lowering costs", 24 April 2019, https://stopillegalfishing.com/press-links/seychelles-testing-drones-to-monitor-waters-for-illegal-fishing-potentially-lowering-costs/.

[60] Scott, "Drones driven by AI will track illegal fishing in African waters", 15 August 2018.

[61] Ibid.

[62] Ibid.

[63] Ibid.

In 2022, the UK supplied the Seychelles Coast Guard with two compact quadcopter drones.[64] These particular drones were purchased in cooperation with the United Nations Office on Drugs and Crime (UNODC).[65] The manner in which these drones have been deployed is unclear. On the whole, Seychelles is emerging as a vibrant maritime security actor in the Indian Ocean and Africa, and its involvement in the U.S.-led Red Sea coalition is proof of that.[66]

6 Nigeria

While Seychelles leads the charge in the Western Indian Ocean (East Coast of Africa), Nigeria takes the lead in West Africa and the Gulf of Guinea. Nigeria is arguably the front-runner in Africa, acquiring, integrating and leveraging drone technology to enhance its maritime security in the Gulf of Guinea. In recent years, the Gulf of Guinea has emerged as a focal point of the international community, largely due to the maritime (in-)security situation in the region that is affecting local economies as well as the supply of oil and gas to the international market. The Gulf of Guinea (in West Africa) is a vast region with a coastline of roughly 6000 kms.[67] This maritime region is delineated from Angola in the south, stretching northwards toward Cameroon, then west across Nigeria, toward Liberia and Sierra Leone.[68] Although incidents of piracy and armed robbery at sea have declined across the region,[69] the Gulf of Guinea faces a number of challenges, including transnational organised crime such as smuggling and trafficking of illicit goods, illicit waste dumping and IUU fishing.[70] The latter, in particular, costs the region's economies approximately $3 billion USD annually.[71]

In terms of naval power and maritime security, Nigeria can be considered one of the most important security guarantors in the Gulf of Guinea. Nigeria's coastline is

[64] defenceWeb, "UK provides drones to Seychelles for maritime surveillance", defenceWeb, 23 February 2022, https://www.defenceweb.co.za/aerospace/uk-provides-drones-to-seychelles-for-maritime-surveillance/.

[65] African Defence Forum, "Eyes in the Sky", 2 February 2023.

[66] Stewart, P. and Davison, J., "US launches Red Sea force as ships reroute to avoid attacks", Reuters, 19 December 2023, https://www.reuters.com/world/middle-east/us-launches-new-multinational-operation-safeguard-red-sea-commerce-2023-12-18/

[67] European External Action Service, EU Maritime Security Factsheet: The Gulf of Guinea (Brussels: EEAS, 2021), https://www.eeas.europa.eu/eeas/eu-maritime-security-factsheet-gulf-guinea_en.

[68] Ibid.

[69] Piracy, as defined by the United Nations Convention on the Law of the Sea (UNCLOS), constitutes 'any illegal acts of violence or detention, or any act of depredation, committed for private ends by the crew or the passengers of a private ship or a private aircraft, and directed: on the high seas, against another ship or aircraft, or persons or property on board such ship or aircraft'. Armed robbery at sea constitutes similar acts of piracy however they occur in a country's territorial waters and EEZ, as opposed to on the high seas (international waters).

[70] Ibid.

[71] African Defence Forum, "Eyes in the Sky", 2 February 2023.

roughly 853 kms, and its EZZ covers a total area of approximately 315,000 square kilometres.[72] Moreover, Nigeria has a vast network of rivers and lakes stretching over 3000 kms onshore, all of which are vital to Nigeria's maritime security and overall national security.[73] With the largest Navy in the Gulf of Guinea,[74] Nigeria is gradually enhancing its monitoring and enforcement capabilities to secure its territorial waters and the parts of the open seas. As maritime security threats, including but not limited to piracy, illegal fishing, and oil bunkering persist, the Nigerian Navy has expanded its drone fleet.[75] Currently, the Nigeria Navy operates nine ADS Aerostar drones, TEKEVER AR3 UAVs and RemoEye 002Bs drones manufactured by South Korea's Uconsystem.[76] In 2019, using both drones and helicopters, the Nigerian Navy conducted surveillance of the Nigerian National Petroleum Corporation (NNPC) pipelines.[77] This forms part and parcel of Nigeria's aim to safeguard its oil industry and maritime trade against piracy, as well as minimise its dependence on private security companies in protecting its oil infrastructure and assets.[78]

In June 2021, Nigeria launched its Integrated National Security and Waterways Protection Infrastructure, also referred to as the Deep Blue Project.[79] The U.S.-backed, $195-million-USD initiative is designed to enhance maritime domain awareness (MDA) and oversee the deployment and coordination of Nigeria's drones and other assets in counterpiracy operations in the Gulf of Guinea.[80] This support is clearly linked to both U.S. and Nigerian national interests. The Deep Blue Project is the first integrated maritime security strategy in both West and Central Africa geared toward combatting piracy, armed robbery at sea and other

[72] Atakpa, S.D., "Marine protected areas and Nigeria's blue economy", *Ships and Ports*, 4 May 2021, https://shipsandports.com.ng/marine-protected-areas-and-nigerias-blue-economy.

[73] Owoputi, E.A. and Oluwatosin, O.O., "Examine the impact of inland waterways transportation on socio-economic development of Ogun State coastal area of Nigeria", FUPRE Journal of Scientific and Industrial Research 3, no. 2 (2019): 57–755; Nze, N.J., Ibe, C.C., Emenike, G.C. and Nze, I.C., "Analysis of operational logistics on the inland waterways of Delta State, Nigeria", Journal of Multidisciplinary Engineering Science and Technology 8, no. 11 (2021): 14,783–14,786.

[74] defenceWeb, "Nigerian Navy critical for African security and stability—USN Commander", *defenceweb*, 13 August 2021, https://www.defenceweb.co.za/featured/nigerian-navy-critical-for-african-security-and-stability-usn-commander/.

[75] Lionel, E., "Nigerian Navy boosts its drone fleet with 12 new Comstrac System Ovation drones", *Military Africa*, 13 December 2023, https://www.military.africa/2023/12/nigerian-navy-boosts-its-drone-fleet-with-12-new-comstrac-system-ovation-drones/.

[76] Ibid.

[77] Ibid.

[78] Ibid.

[79] Babalola, Y., "1 year after Deep Blue launch: Nigeria records 5 months of zero attack on waterways", *Leadership*, 2021, https://leadership.ng/1-year-after-deep-blue-launch-nigeria-records-5-months-of-zero-attack-on-waterways/.

[80] A Anele, K.K., "Analysis of the Enforcement of the Suppression of Piracy and Other Maritime Offences Act in Nigeria: Matters Arising", Criminal Law Forum 34, no. 4 (2023): 375–419, https://doi.org/10.1007/s10609-023-09462-y.

crimes.[81] Through this initiative, Nigeria procured several air assets. In collaboration with its Maritime Administration and Safety Agency (NIMASA), the Deep Blue Project procured a number of shipborne AR3 NET RAY UAS from Portugal's Tekever.[82] The consignment of the Tekever AR3 UAS was delivered at the Apapa Dockyard in Lagos on 16 December 2020.[83] The AR3 UAS are intended to operate in tandem with two Leonardo ATR 42 Maritime Patrol Aircraft of the Nigerian Air Force and two special mission CJ3 Cessna aircraft based in Benin City in the 107 Air Maritime Group (NAF 107 AMG).[84] According to Dr Bashir Jamoh, Director-General of the Nigerian Maritime Administration and Safety Agency, the new systems are expected to provide real-time information to assist in securing Nigeria's coastal areas for critical infrastructure, shipping, and seafarers.[85]

In September 2022, the Nigerian Navy received four AR-500 ship-based drones from the Aviation Industry Corporation of China (AVIC).[86] These drones facilitate the collection of tactical intelligence, wide-area surveillance, identification, target acquisition, precision strikes, as well as maritime relay communication tasks.[87] In addition to the AR-500 drones, China also supplied Nigeria with unmanned combat aerial vehicles (UCAVs), namely two CH-3Bs, two Wing Loong II, and four CH-4Bs.[88] How these systems have been employed and integrated with the rest of the Nigerian defence capabilities are unclear. In 2023, Nigeria acquired twelve Comstrac System Ovation UAVs to improve its maritime security operations.[89] These latest additions are intended to complement the Nigerian Navy's existing air assets to support surface vessels and maritime domain awareness platforms.[90] The advanced capabilities and versatility of these drones are intended to improve "naval surveillance and strategic capabilities".[91]

Nigeria's growing drone order-of-battle, including but not limited to the Comstrac System Ovation, South Korean RemoEye 002B, ADS Aerostar, Portuguese Tekever AR3, Wing Loong II, CH-3B and CH-4B, elevate this West African nation to a position of advantage within its area of responsibly and interest in West Africa. It is worth analysing how the procurement of these systems from various countries factors into cooperation, competitiveness and accompanying geopolitical arrangements in West Africa. It may be argued that this widespread assortment of drones

[81] Babalola, "1 year after Deep Blue launch", 2021.
[82] Oliver, D., "Pioneering unmanned fleet in sub-Saharan Africa", Times Aerospace, 20 March 2024, https://www.timesaerospace.aero/features/defence/pioneering-unmanned-fleet-in-sub-saharan-africa.
[83] Lionel, "Nigerian Navy boosts its drone fleet", 13 December 2023.
[84] Ibid.
[85] Reva and Ramachela, "Can a new wave of drone tech make Africa's seas safer?", 30 November 2022,
[86] Lionel, "Nigerian Navy boosts its drone fleet", 13 December 2023.
[87] Ibid.
[88] Ibid.
[89] Ibid.
[90] Ibid.
[91] Ibid.

fuels a sort of drone race between suppliers (exporters) and procurers (importers). This must be weighed against the prevailing geopolitical fault lines as well as realpolitik dynamics within and amongst regional players in West Africa. Furthermore, it is worth considering whether the procurement of advanced technological systems, in particular maritime capable drones, truly enhances effectiveness or is merely a show of force with no substantive, measurable security gains. Some, among them Babatunde and Abdulsalam,[92] argue that Nigeria does not yet possess the substantial surveillance equipment, specifically drones, required to improve its security. However, this is changing, not only in Nigeria but across West Africa, as more coastal countries acquire drones for maritime security functions. For example, in October 2022, Ghana procured drones to be used by its Boundary Commission and the Ghanaian Navy for maritime patrols and to demarcate its maritime boundary.[93] In 2023, Ghana announced that its Navy had begun acquiring drones with real-time surveillance, coastal radars, and tracking protocols to improve its maritime domain awareness.[94] This will also provide an opportunity for these states to cooperate and provide a regionally integrated unmanned capability purposefully dedicated to the protection of critical infrastructure, maritime trade and people.

7 Drones in the Mediterranean Sea

Whilst drones have been deployed in the Western Indian Ocean, the Horn of Africa and the Gulf of Guinea, they have also been utilised in the Mediterranean Sea. In addition to Seychelles and Nigeria, other African states are steadily procuring and deploying maritime drones. Some of the more recent acquisitions and deployments have been in North Africa on the coastline of the Southern Mediterranean Sea. In December 2022, Morocco acquired at least four SeaGuardian MQ-9B aerial drones, manufactured by the US General Atomics Aeronautical Systems.[95] These drones have a range of 6000 nautical miles (11,100 kms) and can survey vast areas of land and sea.[96] According to Adu-Gyamfi et al.,[97] Morocco deployed drones to track illicit maritime activities, including oil spills.

[92] Babatunde, E.O. and Abdulsalam, M.M., "Towards maintaining peacefulness of the sea: legal regime governing maritime safety and security in Nigeria", Beijing Law Review 12, no. 4 (2021): 529–559.

[93] Africa Defence Forum, "The Dizzying Possibilities of Drones in Africa", 20 October 2023.

[94] Modern Ghana, "We're equipping navy with drones and radars to improve maritime security—Akufo-Addo", *Modern Ghana*, 29 August 2023, https://www.modernghana.com/news/1255127/were-equipping-navy-with-drones-and-radars-to.html.

[95] Dorgambide, M., "Morocco receives state-of-the-art Israeli drones", Atalayar, 24 June 2023, https://www.atalayar.com/en/articulo/new-technologies-innovation/morocco-receives-state-of-the-art-israeli-drones/20230620170237186987.html.

[96] Ibid.

[97] Adu-Gyamfi, S., Gyasi, R.M. and Darkwa, B.D., "Historicizing medical drones in Africa: a focus on Ghana", History of Science and Technology 11, no. 1 (2021): 103–124.

Although not under the command of African countries, drones have been deployed by European countries in the Mediterranean Sea to detect and monitor the movement of migrants and refugees from North Africa to Europe[98]—which co-incidentally also provide opportunities for transnational crime syndicates to engage in human trafficking. These drones are equipped with advanced capabilities; during the day, the drones use cameras to search for vessels and at night, they employ infrared sensors to detect vessels.[99] The use of drones by European coastal states bordering the Mediterranean Sea, including Greece, Italy and Portugal, and the European Border and Coast Guard Agency, also known as Frontex,[100] is gaining momentum. Since 2016, as part of an Aerial surveillance Service, Frontex has been using drones to monitor parts of the Mediterranean Sea,[101] particular areas identified as being the main transit (and smuggling) routes. To facilitate this process, the European Commission has adopted several laws and regulations. The new Frontex Regulation 2019/1896 extends the reach and operational scope of Frontex far into African waters. In no uncertain terms, it is clear that the new regulation will allow Frontex to "become more active outside the European Union with the ability to conduct operations in non-EU countries that do not neighbour the EU".[102] In terms of this regulation, pursuant to the consent of all parties involved and working agreements with the relevant government, Frontex can operate drones in the eastern and central Mediterranean within a radius of approximately 250 nautical miles (463 kms) from the point of their departure.[103] Under this regulation, Frontex can conduct reconnaissance in the waters near Egypt, Libya and Tunisia.[104] Essentially, Frontex can deploy its drones once the relevant countries have authorised (granted permission) the missions. It may be worth exploring how such bilateral agreements between North African states and the EU are perceived by the African Union (UN).

Evidence suggests the EU Naval Force in the Mediterranean (EUNAVFOR MED), also known as Operation Irini cooperates more extensively with the Libyan

---

[98] Klein, N., "Maritime autonomous vehicles and international laws on boat migration: lessons from the use of drones in the Mediterranean", Marine Policy 127, no. 1 (2021): 1–10, https://doi.org/10.1016/j.marpol.2021.104447.

[99] Ibid., 4.

[100] Frontex, "New Frontex Regulation comes into force", *Frontex*, 4 December 2019, https://www.frontex.europa.eu/media-centre/news/news-release/new-frontex-regulation-comes-into-force-S0luwe.

[101] Monroy, M., "Europe's drone agency: Frontex paves the way for military drones in domestic airspace", *Journalismus Von Links*, 3 October 2023, https://www.nd-aktuell.de/artikel/1176711.festung-europa-frontex-europas-drohnenagentur.html.

[102] Frontex, "New Frontex Regulation comes into force", *Frontex*, 4 December 2019, https://www.frontex.europa.eu/media-centre/news/news-release/new-frontex-regulation-comes-into-force-S0luwe.

[103] Mazzeo, A., "Border surveillance, drones and militarisation of the Mediterranean", *State Watch*, 6 May 2021, https://www.statewatch.org/analyses/2021/border-surveillance-drones-and-militarisation-of-the-mediterranean/.

[104] Ibid.

maritime authorities, specifically by sharing information collected using the Italian Ministry Predator drones with the Libyan Coast Guard.[105] Securing a solid working agreement with Libyan authorities may be difficult, given the various warring parties and their contested claims of authority. This much was made clear in late 2019 after the new Frontex Regulation came into force; on 20 November 2019, an Italian Air Force MQ9 Reaper Predator drone crashed in Libya, after it was shot down by a missile of the Libyan National Army (LNA).[106] In response to this, the Italian Ministry of Defence stated that the drone was being used to "support and bolster the Libyan Coast Guard and Navy to counter unlawful maritime trafficking".[107]

There remain obstacles to the enhanced effective collaborative use of drones in the Mediterranean Sea. Despite the benefits of minimal cost, enhanced safety and efficiency in drone deployment, the use of drones to curb 'migrant smuggling' has stirred some controversy. Human rights advocates, including Amnesty International,[108] have confirmed that data collected by the drones, namely the precise location of migrant boats, is relayed to Libyan authorities to intercept vessels and return the passengers to Libya, where they will face human rights abuses.[109] This type of utilisation of drones is inherently flawed and in direct violation of international human rights law and international humanitarian law. Analysts, including but not limited to Klein,[110]argue that the deployment of drones for surveillance and the absence of surface assets to rescue migrants in distress skirts the European Union's international legal and humanitarian obligations. In accordance with the United Nations Convention on the Law of the Sea (UNCLOS), all states have a responsibility to rescue persons in distress. Article 98 '*Duty to Render Assistance*' stipulates that "Every State shall require the master of a ship flying its flag, in so far as he can do so without serious danger to the ship, the crew or the passengers: to render assistance to any person found at sea in danger of being lost; to proceed with all possible speed to the rescue of persons in distress, if informed of their need of assistance, in so far as such action may reasonably be expected of him'.[111] Essentially, if a littoral state employs drones to monitor maritime spaces and, in the process, identifies persons in distress but fails to or avoids deploying rescue vessels, that state is in violation of UNCLOS.

---

[105] Ibid.

[106] Ibid.

[107] Ibid.

[108] Amnesty International, "EU: Frontex's lack of transparency on Libyan cooperation to be heard in court", 10 October 2023, https://www.amnesty.org/en/latest/news/2023/10/eu-frontexs-lack-of-transparency-on-libyan-cooperation-to-be-heard-in-court/.

[109] Klein, "Maritime autonomous vehicles and international laws on boat migration", 2021, 5; Mazzeo, "Border surveillance, drones and militarisation of the Mediterranean", 2021, 15.

[110] Klein, "Maritime autonomous vehicles and international laws on boat migration", 2021, 2.

[111] United Nations, United Nations Convention on the Law of the Sea (New York: United Nations, 1982), https://www.un.org/depts/los/convention_agreements/texts/unclos/unclos_e.pdf.

Klein[112] confirms that the deployment of drones in the Mediterranean Sea has occurred alongside the decline of surface assets and resources required for rescue efforts. The deployment of drones by European authorities, without surface vessels to rescue persons in danger, raises ethical questions about the appropriateness and legality of drones, specifically in the maritime domain, and vis-à-vis irregular migration. The EU's concerns about migrant smuggling, maritime security and national security are understandable. However, its security concerns cannot be at the expense of human lives. Notwithstanding, it seems that this trend of drone deployment to detect and deter irregular migration and other transnational crimes will continue. Owing to the utility of drones for surveillance, information sharing, border security and monitoring migration-smuggling routes, Klein[113] suggests that European states might provide drones to origin and transit countries, like Libya, in the foreseeable future.

8 Non-state Actor Drone Use in the Maritime Domain

It is not only state actors deploying drones at sea, but also benevolent non-state actors such as non-governmental organisations (NGOs) and humanitarian and aid workers. The Migrant Offshore Aid Station (MOAS) has deployed drones to conduct search and rescue efforts alongside a fully crewed and equipped vessel to rescue migrant boats in distress.[114] Another NGO called Open Arms, has employed drones to "track the trajectory of ships to determine if they are drifting and in distress" and subsequently send the information to the rescue vessels.[115] The reality is that if benevolent non-state actors can deploy drones at sea, so can more nefarious actors. Klein[116] states that criminal groups, such as maritime smugglers, may use drones to monitor the authorities' activities to determine how and when to make illicit crossings without detection.

While drones can be a force for good to promote maritime security, they can also threaten maritime security. There is already an observably growing trend of non-state actors, such as criminal syndicates and even armed militant groups, using drones in their day-to-day operations. There have been incidents of armed groups using drones in the northwest Indian Ocean. For example, the Yemeni Houthis have used drones to conduct a handful of attacks in the Red Sea.[117] Between January 2017 and June 2021, the Houthis carried out 24 attacks against commercial

[112] Klein, "Maritime autonomous vehicles and international laws on boat migration", 2021, 7.
[113] Ibid., 6.
[114] Ibid., 2.
[115] Ibid.
[116] Ibid.
[117] Reva and Ramachela, "Can a new wave of drone tech make Africa's seas safer?", 30 November 2022; Al Jazeera, "French frigate shoots down drones over Red Sea: military", *Al Jazeera*, 10 December 2023, https://www.aljazeera.com/news/2023/12/10/french-frigate-shoots-down-drones-over-red-sea-military.

vessels, port infrastructure, and Saudi oil production and distribution facilities.[118] There are reports of drones being used for landward surveillance by non-state actors, such as armed groups in the Democratic Republic of Congo (DRC) and insurgents in Cabo Delgado in Mozambique.[119] As far as is known, such activities have been restricted to land; however, given the relative ease of procurement and deployment, it is perhaps only a matter of time before African non-state actors begin leveraging the benefits of maritime drones, as the Houthis have. The proliferation of drone technology in Africa's maritime environment and its use by both state actors and, in particular, non-state actors warrant further reflection. The trend of non-state actors employing drones for nefarious reasons requires greater reflection on the potential benefits of anti-drone technology or counter-unmanned aerial/aircraft systems (C-UAS). C-UAS technology includes portable and fixed systems such as radar systems, radio frequency jamming devices, drone detection sensors, and drone capture nets. These systems monitor and detect potential drone threats and collect data and digital evidence. Presently, some African countries, among them Morocco, Ethiopia, Zambia, Cote d'Ivoire, Mozambique, Algeria, Libya, Angola, Egypt and Rwanda, have active C-UAS solutions. However, much of this technology is sourced externally from companies based in the US, the UK, China, Israel and even Poland. Going forward, it would be worth not only conducting empirical research on the merits of these systems but also practically integrating these systems and testing their utility in Africa's maritime environment.

9 Future of Drones in the African Maritime Domain: Challenges and Opportunities

The future of drones in Africa's maritime domain is somewhat uncertain. The rapid deployment of drones has engendered new challenges. Despite the benefits of drones, the technology is imperfect, and many factors may delay deployment and subsequent integration into defence capabilities and doctrine. There are many challenges vis-à-vis drone deployment in the maritime domain. For Colin,[120] these issues include autonomy, cybersecurity and connectivity. First, the size of the drone determines its tasks, its range, its speed, the length of time it can operate out at sea, and its vulnerability to weather and the environment. The issue with smaller

[118] Allworth, *The South African Air Force (SAAF), unmanned aircraft systems and national security*, 41; Bueger, C. and Stockbruegger, J., "Maritime security and the Western Indian Ocean's militarisation dilemma", African Security Review 31, no. 2 (2022): 195–210, https://doi.org/10.1080/10246029.2022.2053556.

[119] Allen, K., "Drones in the hands of insurgents: how Africa can prepare", *Institute for Security Studies*, 4 November 2022, https://issafrica.org/iss-today/drones-in-the-hands-of-insurgents-how-africa-can-prepare.

[120] Colin, L., "Military drones: current trends in the market", Polytechnique Insights: A Review by Institut Polytechnique de Paris, 2022, https://www.polytechnique-insights.com/en/columns/geopolitics/military-drones-current-trends-in-the-market/.

systems is their limited operational scope, functions and range, which translates into a low return on investment. Larger platforms can carry out a wider variety of tasks for longer periods of time; however, this comes at a much greater cost. While a small vessel can be purchased for around $350,000 USD, a bigger one could cost $2 million to $5 million USD.[121] Despite the higher costs, larger vessels are better suited for maritime security at an operational cost similar to that of their smaller counterparts. Larger systems can operate in the open ocean, carry high-bandwidth communication systems, and deploy powerful and stable payload sensors.[122]

The cost–benefit of larger unmanned systems becomes apparent over time compared to traditional naval capabilities. Large unmanned systems provide a scalable way to achieve the geographic and temporal coverage that is increasingly required in today's maritime environment.[123] However, costs will still be incurred. Drone licensing fees are somewhat exorbitant compared to what local (African) start-ups can afford.[124] Another financial challenge is the cost and time needed to train drone operators. Some African authorities may be reluctant to spend additional capital to train their operators and would rather opt to have these operators learn on the job. Some analysts have argued that this "do-it-yourself" approach may result in idle, rarely used technology or drone wreckages in the oceans.[125] The risk of potential cyberattacks is linked to the issue of ill-trained operators and vulnerable systems. Drones rely on satellite navigation and data, which is susceptible to disruptions. Thus, drones, as well as the data they collect, can be hacked. This makes GPS spoofing a significant concern.[126] Both state and non-state actors can interfere with the GPS to manipulate and alter its location coordinates. In a worst-case scenario, the spoofing of a drone could be considered an act of war or result in major legal ramifications.[127] Despite the threat of cyberattacks, the endurance, sensors and payload capacity of drone technology are rapidly improving.[128] Even so, changes and improvements in the laws regulating drone use in the maritime domain are necessary.

According to an AU Report titled Drones of the Horizon: Transforming African Agriculture, in 2017, only 14 African countries had dedicated drone regulations.[129]

[121] Reva and Ramachela, "Can a new wave of drone tech make Africa's seas safer?", 30 November 2022.
[122] Ibid.
[123] Ibid.
[124] Scott, "Drones driven by AI will track illegal fishing in African waters", 15 August 2018.
[125] Africa Defence Forum, "The Dizzying Possibilities of Drones in Africa", 20 October 2023.
[126] Reva and Ramachela, "Can a new wave of drone tech make Africa's seas safer?", 30 November 2022.
[127] Ibid.
[128] Rawley and Patmon, "Drones in Africa: a leap ahead for maritime security", 9 October 2018.
[129] African Union, Drones on the Horizon: Transforming Africa's Agriculture (Addis Ababa: African Union, 2018), https://www.nepad.org/publication/drones-horizon-transforming-africas-agriculture.

Among the countries leading the charge in drone regulation is South Africa.[130] However, there remains no aviation regulatory body on the African continent, nor is there an overarching regulatory framework.[131] Therefore, in their capacity, each state regulates drones as they see fit. The deployment of drones and other unmanned systems is constrained by relatively outdated and restrictive regulations, namely the disconnect between prevailing directives and the requirements for using unmanned systems in African waters.[132] For example, cameras or sensors onboard unmanned vessels may detect a ship's presence in restricted waters; however, that alone may not be sufficient to arrest and prosecute individuals. This raises issues of both digital connectivity and policy implementation and harmonisation. Nyman[133] confirms this, noting that the deployment of drones for monitoring does not resolve issues of enforcement vis-à-vis ocean governance, and it remains difficult to ascertain the impact of maritime crimes.

Another challenge that is regulatory in nature is the issue of the legal loopholes surrounding the collection of data (evidence) and the delineation of jurisdictions.[134] The scenario below illustrates the complexity of regulation:

***The Scenario*:**

A privately owned, licensed vessel is identified fishing in a marine protected area in the territorial waters of Country A.

This vessel, operated by another private company, is sailing under a "flag of convenience" (when a company uses open registries to fish in waters beyond the jurisdiction of coastal nations). The flag belongs to Country B.

The vessel's crew are of multiple nationalities from Countries C, D and E.

A drone belonging to Country F surveys the vessel's destructive fishing practices (trawling) in Country A's territorial waters.

***The questions*:**

Who has the onus and authority to detain and prosecute the guilty parties, and who are said guilty parties?

Is the guilty party the owner of the vessel, or is it the operators of the vessel, or it is the captain and crew (potential victims of human trafficking and forced labour) caught in the act of illegal fishing?

---

[130] Zoldi, D.M.K., "African drone regulations: how easy is it to operate in the African market?", Drone Life, 20 May 2023, https://dronelife.com/2021/05/20/african-drone-regulations-how-easy-is-it-to-operate-in-the-african-market/.

[131] Ibid.

[132] Reva and Ramachela, "Can a new wave of drone tech make Africa's seas safer?", 30 November 2022.

[133] Nyman, E., "Techno-optimism and ocean governance: New trends in maritime monitoring", *Marine Policy* 99 (2019): 30–33, https://doi.org/10.1016/j.marpol.2018.10.027.

[134] Reva and Ramachela, "Can a new wave of drone tech make Africa's seas safer?", 30 November 2022.

Given the infringement occurred in Country A's territory, does Country A have the jurisdiction and authority to detain and prosecute whomever it deems guilty? Is country B liable because the vessel is flying its flag?

Clearly, an urgent need remains to align and harmonise drone laws across the continent. They may not necessarily be uniform; however, some sort of common framework is essential. Governments should consider amending regulations for drone operations and ensure that data collected can be used to make arrests and proceed to trials. This is a vital step toward fully integrating drone technology into maritime security operations. Although unmanned systems cannot yet fully supplement traditional maritime security forces, they can potentially serve as force multipliers, supplementing and supporting the role of traditional naval operations and disrupting and deterring illegal activity at sea.[135] In this sense, drones play a complementary role. However, the current outdated regulations delay and complicate the deployment of drones, which must be remedied.

In the future, given the utility of drones, African countries should consider procuring other autonomous vehicles, such as uncrewed surface vessels (USVs), which have already proven useful against larger vessels. As such, the Yemeni Houthis have deployed similar drones, specifically explosive boats with a single pilot on board or remote-controlled, in their attacks against oil tankers in the Red Sea.[136] The potential of USVs to do harm means that there is also potential to do good. Another innovation worth considering is underwater unmanned vehicles (UUVs). UUVs, also called remotely operated vehicles (ROVs), are essentially unmanned remotely controlled submarines; they are waterproof submersibles capable of exploring the marine environment.[137] Allworth[138] states that a combination of both UAS and UUVs may improve maritime situational awareness (MSA). Currently, non-state actors are using homemade UUVs to transport narcotics by sea. In July 2022, Spanish authorities seized three underwater drones capable of carrying roughly 200 kg of cargo.[139] These underwater drones, labelled 'narco-subs', are gaining traction as a preferred mode of drug trafficking. Arguably, these submersibles portend a major maritime security threat; that is, if such vehicles are capable of transiting vast maritime spaces to transport narcotics, they may also be

[135] Dikmen, M., Atalay, M. and Gümüş, B., "Role of unmanned aircraft systems in maritime security", *American Scientific Research Journal for Engineering, Technology, and Sciences* 26, no. 3 (2016), 164–171.

[136] Sutton, H.I., "Disguised explosive boat may be new threat to tankers off Yemen", *Forbes*, 4 March 2020, https://www.forbes.com/sites/hisutton/2020/03/04/new-disguised-explosive-boat-may-threaten-tankers-off-yemen/?sh=288c62371ad2.

[137] Van Nieuwkerk, A., "The Impact of Digital-Driven Warfare on Africa", The Thinker 90, no. 1 (2022): 20–27.

[138] Allworth, *The South African Air Force (SAAF), unmanned aircraft systems and national security*, 42.

[139] Jones, S., "Spanish police seize underwater drones designed to carry drugs", The *Guardian*, 4 July 2022, https://www.theguardian.com/world/2022/jul/04/spanish-police-seize-underwater-drones-designed-to-carry-drugs.

able to transport more dangerous goods or substances. Theoretically, an underwater drone loaded with explosives may cause major damage should it collide or be detonated remotely near a passing vessel, beneath an oil rig or in a harbour. With this kind of threat looming beneath the surface and the rapidly evolving innovations of criminal actors, the necessity for African maritime authorities to anticipate and mitigate such threats cannot be understated.

The stockpiling of drones by some African countries perhaps signals a rapidly developing arms race, or more aptly 'drone race'. This raises questions about the geopolitics of drones, competition and possible tension. Dyer[140] argues that drones may promote maritime security, but their mere presence can also exacerbate prevailing conflicts and geopolitical tensions. According to Prasanga,[141] understanding the "strategic impact of drone proliferation" is crucial for "assessing regional power balances, potential conflict, and cooperative security efforts". While there appears to be a race to acquire as many drones as possible, there is also a race amongst suppliers to set themselves apart and offer appealing deals to African countries. Leading this charge is the US, Israel in North Africa, and recently, Turkey, as part of its Turkish Drone Diplomacy, where it is actively marketing its drone technology in Africa.[142] However, there might be unintended consequences of this drone proliferation in Africa. According to Bueger and Stockbruegger,[143] policymakers have growing concerns about "hybrid warfare and maritime greyzone attacks" on shipping lanes, including maritime drone attacks. To date, no African country has used drones to execute air strikes in African waters. However, that may change in the future as Africa's maritime domain becomes increasingly militarised and rife with insecurity. Some analysts argue that Africa may become a "testing ground" for new phases and prototypes of drone technology, including the incorporation of AI.[144] This should be a warning to African countries and a signal to bolster maritime capabilities, technological capacity and policy and regulatory frameworks.

There are some further research areas and questions worth exploring. For example, does the use of drones in the maritime domain inevitably lead to or equal militarisation, even when said drones are non-combative and focus more on nontraditional maritime security issues such as illegal fishing? The latter is especially important concerning African coastal communities and the blue economy. It is crucial that we explore the benefits and implications of using drones to combat IUU fishing. Another area worth exploring is the use of drones for recreational

---

[140] Dyer, *Defending African Maritime Sovereignty with Finite Resources, 2018,* 25.

[141] Prasanga, "The strategic impact of military drone proliferation on Indo-Pacific maritime security", 23 November 2023.

[142] Vreÿ, F. and Blaine, M., "Exploring the Maritime Nexus in Turkey's African Agenda", Journal of Balkan and Near Eastern Studies 26, no. 3 (2024), 346–359, https://doi.org/10.1080/19448953.2023.2236513.

[143] Bueger & Stockbruegger, "Maritime security and the Western Indian Ocean's militarisation dilemma", 2022, 198.

[144] African Defence Forum, "The dizzying possibilities of drones in Africa" 20 October 2023.

fishing. Some individuals in South Africa have begun using drones to search for fertile fishing grounds and generally to fish in hard-to-reach areas.[145] This raises issues of safety, legality and even environmental questions. In terms of marine conservation and environmental protection, it is necessary to investigate the kinds of effects drone fishing may have on fish stocks and endangered species, and the risks posed to other animals such as marine mammals and birds. A drone obscured from the pilot's line of sight, or even drone debris, may have major implications for ecological systems. Whilst drone fishing is accessible to affluent individuals, it arguably disadvantages local small-scale fishers who lack the financial resources to acquire and deploy commercial drones. This may exacerbate competition between local fishers in South Africa's coastal communities.[146] The advent of drone fishing may be a positive development, granted it is practised equitably. Until then, this emerging frontier requires specialised interrogation and the implementation of the appropriate regulations. Linked to this is the aspect of AI, its integration in maritime drone technology and how this may aid and/or complicate operations. Over and above, drone laws must be established and/or amended, and regulations around AI and its integration are also necessary.

10 Conclusion

The nature of Africa's maritime domain is rapidly evolving. There are an ever-growing number of emerging challenges and trends. Alongside the new developments in the maritime environment, Africa's drone market, both military and commercial is growing exponentially, in part due to the growing demand for drones. The lack of capability and capacity of several African maritime security forces and the prospect of gaining a competitive advantage make the continent a highly profitable market for maritime drone suppliers. Thus far, seemingly and predominantly, Africa is still a consumer rather than a producer in the development and deployment value chain of drones. Consequently, African countries are sometimes inadvertently participants in external alliances and geopolitical squabbles. In and of itself, Africa's dependence on extra-regional actors for the provision and operation of maritime drones is an inherent technological (cyber) vulnerability and risk to autonomy and integrity.

At the moment the ability to manufacture highly advanced maritime-capable drones may be beyond the capacity of individual states, except for South Africa and Morocco. However, it is worth exploring the possibility of maritime drone production at the regional level where countries can pool their resources and share expertise. This may in turn create employment opportunities, enhance integration

[145] Winkler, A.C., "Drone fishing in South Africa is a danger to sharks and may be unfair to other fishers—study", *The Conversation*, 30 November 2023, https://theconversation.com/drone-fishing-in-south-africa-is-a-danger-to-sharks-and-may-be-unfair-to-other-fishers-study-218267#:~:text=In%20addition%20to%20polluting%20the,catch%20for%20food%20or%20income.
[146] Ibid.

and bolster interoperability. Whilst the rapid deployment of maritime drones is revolutionising the African maritime environment, it also complicates ocean governance, specifically in terms of regulatory frameworks, and harmonising policies. This must be addressed swiftly and collectively amongst African countries, at the continental level and within the various sub-regional bodies.

Over and above, there remains a gap in the literature on drone deployment for maritime security and safety in Africa. This chapter aimed to bridge this gap, however, there is still a need to conduct in-depth research. There are some questions and areas warranting further investigation namely, how does the procurement of maritime-capable drones from several suppliers, impact path dependence, trajectories and the independence of foreign and security policy of African states. Despite having no combat weapons delivery functions and serving primarily as a means of security and development, does the mere proliferation of drones reflect a sort of militarisation of Africa's maritime domain? Alternatively, is this dependent on how and for what purposes maritime drones are deployed? Could African authorities, namely navies begin launching drone strikes at sea against perceived threats and what would be the implications? Besides safety and security, will coastal communities, such as fishing villages equitably benefit from the development potential offered by maritime drone deployment?

Technology in general, and drones in particular, can be a force for good, it all depends on who is using the tech and for what purpose. Where utilised for humanitarian reasons, maritime safety, security and other development-related aspects, drones can reinforce efficiency and bolster resilience. However, when used for harmful and nefarious reasons, by individuals or entities with ill intent, drones have the potential to undermine vital national security interests. Drone technology is and will continue to shape the African maritime domain, policies, strategic objectives as well as operational and legal frameworks. More empirical studies on maritime drone deployment in Africa are required to fully explore the intricacies of this rapidly evolving field and its associative implications.

Tshegofatso Johanna Ramachela holds an MA degree in International Relations from the University of Pretoria, South Africa. Previously, she was a consultant for the Maritime Project at the Institute for Security Studies (ISS). Ramachela is currently a Research Associate for the Ocean Regions Programme. She is pursuing a master's degree in Security, Intelligence, and Strategic Studies (IMSISS) at the University of Glasgow and its partner institutions. Her key thematic research areas include maritime crimes, such as narcotics and arms trafficking at sea, the use of maritime (semi-)autonomous vehicles in Africa's maritime domain, and risk-based border management within the context of national security.

Unmanned Aerial Systems and Violent Non-state Actors in Africa: Proliferation, Adaptation and Use

Karen Allen

Abstract

This chapter examines the proliferation, adaptation, and utilization of Unmanned Aerial Systems (UAS) by violent non-state actors (VNSAs) throughout Africa, a phenomenon influenced by the democratization of technology and the spread of network affiliations. Initially exemplified by the Islamic State of Iraq and Syria (ISIS) in the Middle East, UAS technology has demonstrated versatility for deployment in intelligence, surveillance, reconnaissance, and limited kinetic operations by armed groups within Africa. Although the documented malicious use of UAS in Africa is comparatively less widespread than in other regions, factors such as open-source knowledge, network connections, and the accessibility of technology pose increasing security concerns. This chapter undertakes a comparative analysis of the potential and actual applications of UAS by African VNSAs in contract to military and civilian UAS usage, highlighting the shifting threat landscape and the pressing necessity for coordinated policy responses. There is a focused emphasis on the dual-use nature of UAS technology, which, despite its economic and humanitarian advantages, may empower non-state actors to exert territorial influence and psychological pressure beyond their traditional capabilities. The chapter advocates for proactive engagement by regional organizations to develop countermeasures and regulations, drawing insights from globally acquired lessons.

K. Allen (✉)
Department of War Studies, King's College, London, UK
e-mail: karen@karenalleninternational.com

Centre for Information Integrity in Africa, University of Stellenbosch, Stellenbosch, South Africa

W. S. Coetzee and D. Putter (eds.), *Drones in the African Battlespaces*, Southern Space Studies, https://doi.org/10.1007/978-3-031-87147-4_8

1 Introduction

In 2016, the full potential of weaponising and adapting unmanned aerial systems (UAS) became apparent. In Iraq and later in Syria, ISIS demonstrated an impressive ability to harness new technologies and use them to enhance their strategic footprint. There has been much focus on the regular military use of drones and its alarming consequences for international humanitarian law. However, there had been remarkably little interest among UN agencies and other policy bodies, until recently, in the use of UAS by violent non-state actors such as armed groups in Africa. Consequently, there has been very little funding for research.

I attended a conference at the Stockholm Peace Research Institute in 2020 where ICRC's Chief Legal Officer, Cordula Droege, was joined on the platform by a French General, fresh from the United Nations Multidimensional Integrated Stabilisation Mission in Mali (MINUSMA). They spoke of the innovative character of insurgent groups and their impact on international humanitarian law—the laws of armed conflict were originally conceived at a time when regular wars dominated the landscape rather than the asymmetric wars that we see today. They spoke of how insurgent groups—most notably Daesh or ISIS and al-Qaeda and its affiliates—had developed improvised explosive devices (IEDs), graduating from tripwire technology to devices detonated by mobile phones. These became a hallmark of the second Iraq war, with Action on Armed Violence noting that with the exception of 2017, "IEDs have been responsible for more civilian deaths than any other explosive weapon type in each and every year over the last decade".[1] At the Stockholm meeting, a parallel started to be drawn between IED technology and a relatively new, affordable, and easily accessible technology: UAS, colloquially referred to as drones. It felt like only a matter of time before one would observe similar uses identified in the Middle East play out across Africa.

While the documented scale of malevolent UAS use by non-state armed groups across Africa does not appear to have been as extensive as in the Middle East, there has been limited diffusion of the technology across the African continent. Some scholars, such as Chávez and Swed, posit that network affiliations are some of the greatest indicators of the likelihood of armed groups' adoption of drone technology.[2] As ISIS and al-Qaeda affiliates have spread their networks into Africa, most notably the Sahel and the Horn of Africa, the exchange of technological know-how and real-world use cases have proliferated. Moreover, open-source knowledge from the internet in the form of manuals, chat forums and dark web sources, Chávez and Swed argue, has also democratised the spread of technological knowledge to armed groups. Furthermore conflicts such as Russia-Ukraine which have seen the

[1] Overton, I, "A decade of global IED harm reviewed", *Action on Armed Violence*, 15 October 2020, https://aoav.org.uk/2020/a-decade-of-global-ied-harm-reviewed/.

[2] Chávez, K. and Swed, O., "The Proliferation of Drones to Violent Nonstate Actors", *Defence Studies* 21, no. 1 (2021), 1–24, https://doi.org/10.1080/14702436.2020.1848426.

arming of drones by users in their garages at home, has arguably served as an inspiration to others globally.[3]

This chapter investigates the adaptation, use and diffusion of UAS technology by non-state armed groups in Africa. This stands in contrast to investigations which have focused on weaponised UAS supplied by friendly states, such as those which have reportedly been supplied to Houthi rebels in Yemen by proxies such as Iran.[4] It sets out the context by briefly touching upon the rapidly expanding drone ecosystem in Africa, noting that, by and large, the technology has had positive applications in the fields of economic development, healthcare delivery, agriculture, climate change forecasting and disaster response.[5] It also charts the evolution of military drone technology in Africa and its use in the fight against terrorism. It then reflects on the weaponisation of off-the-shelf hobbyist UAS technology by armed groups in Iraq, noting that the Battle of Mosul provided a theatre in which to observe the use of drones for surveillance, target identification and as a weapon deployment system by ISIS. West Point scholar Don Rassler's detailed observations and analysis of ISIS use of UAS in Iraq shed light on a phenomenon in what was arguably the most detailed study of the time. His research begged the question: Could similar use be observed in other settings? In many ways, this chapter seeks to address that question by reflecting on the various uses of UAS by non-state armed groups in Africa.

While differences in geography, topography, politics, and conventional weapons proliferation on the African continent may shape the future in this regard, there have been some similarities with the Middle East in the use of UAS for reconnaissance, surveillance and propaganda purposes. There is only limited documentation of instances of kinetic use. This may be due to limited research. Furthermore, with more documented examples of hobbyist drones being used in other arenas like the Russia-Ukraine and Israel-Palestine conflicts, one may assume that ongoing monitoring in Africa in the future may surface any trends that may have been inspired by the uptake of hobbyist drones elsewhere. This chapter examines the range of uses of UAS by violent armed groups in Africa and notes the potential for surprise attacks or intelligence gathering, both inside and outside of zones of conflict. This enables malevolent actors to hold power over a larger area than they might otherwise control. It also arms them with a psychological threat over populations simply by having access to a tool they could potentially deploy.

While bodies such as the African Union seek to highlight the manifold economic benefits that drone technology brings, understanding the threats from an

---

[3] Rassler D, "On the Horizon: The Ukraine War and the Evolving Threat of Drone Terrorism", *CTS Sentinel* 18, no. 3, March 2025. https://ctc.westpoint.edu/on-the-horizon-the-ukraine-war-and-the-evolving-threat-of-drone-terrorism/

[4] Conflict Armament Research Ltd., "Evolution of UAVs employed by Houthi forces in Yemen", *Storymaps*, 2020, https://storymaps.arcgis.com/stories/46283842630243379f0504ece90a821f.

[5] APET Secretariat, "A Bird's-Eye View: The Application of Drone Technology in Rapid Disaster Response Management", AUDA-NEPAD, 11 February 2021, www.nepad.org/blog/birds-eye-view-application-of-drone-technology-rapid-disaster-response-management.

African perspective is critical for the continent to energetically engage in discussions like those spearheaded by the Global Counterterrorism Forum (GCTF) to operationalise the Berlin Memorandum on Good Practices for Countering Terrorist Use of Unmanned Aerial Systems.

2 Drones Come to Africa

The first documented use of an UAS for drone strikes for lethal purposes in Africa was approved by the Obama administration on 23 June 2011. The US President sanctioned the use of a Predator drone (a favoured US military aerial delivery platform) in Somalia. The restive Horn of Africa territory had always been something of an Achilles heel for the US military after its retreat from Somalia following the Black Hawk Down incident nearly two decades earlier.[6] That episode dramatically changed the US appetite for deploying boots on the ground in the region. For President Obama, his personal standing would continue to be dented by a reflex to use uncrewed air power to keep US service personnel safe. As one news report trumpeted during an assessment of the Obama administration, there were ten times more air strikes in the covert war on terror under President Obama compared to his predecessor, George W. Bush.[7] The headlines spoke not only of terrorist targets but also of the extensive collateral damage of drone warfare—both physical and political—caused by the high counts of civilian casualties. Furthermore, the risks of breaching international humanitarian law's fundamental principles of distinction, proportionality and necessity have led to an energetic debate about the ethics of using drones in battle and concerns that their (at times) indiscriminate nature may serve to radicalise communities who might otherwise be vital sources of human intelligence.[8]

The targets of the 2011 Predator drone strike in Somalia were reportedly al-Shabaab militants who were associates of Anwar al-Awlaki, the leader of the Yemen-based al-Qaeda affiliate in the Arabian Peninsula. The mission, which sought to erode the strategic and operation ties between islamist militants in both Somalia and Yemen, received widespread publicity at the time and arguably dramatically altered the course of counterterrorism and counter-insurgency operations globally in the years that followed.

The Somalia operation signalled a rapid escalation in what was dubbed a 'drone war' driven by a desire to curb the rise of islamist-inspired terrorism, as well as

[6] NPR, "What a Downed Black Hawk in Somalia Taught America", 5 October 2013, www.npr.org/2013/10/05/229561805/what-a-downed-black-hawk-in-somalia-taught-america.

[7] Ibid.

[8] Emmerson, B., "Statement by Ben Emmerson, Special Rapporteur on the Promotion and Protection of Human Rights and Fundamental Freedoms While Countering Terrorism at the 25th session of the Human Rights Council", United Nations Human Rights Office of the High Commissioner, 11 March 2014, www.ohchr.org/en/statements/2014/07/statement-ben-emmerson-special-rapporteur-promotion-and-protection-human-rights.

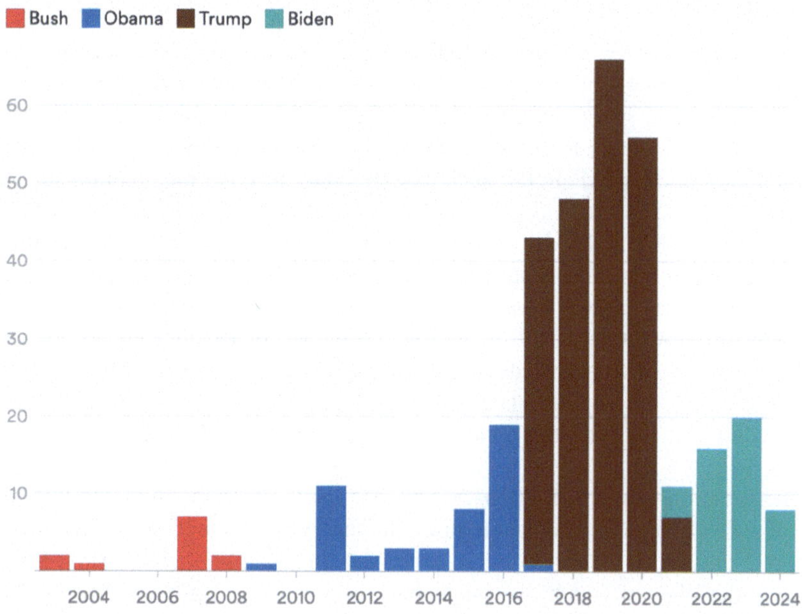

Fig. 1 U.S. air, drone, and ground operations in Somalia, by administration and year. *Source* New America[10]

what appeared to be a deliberate strategy by the administration to demonstrate to Western capitals that they were taking action against terrorists while mitigating the effects of US servicemen and women being killed in battle. It opened a lively debate in the media about the limits to drone warfare. This incident also led to much introspection among defence departments outside the US about how to mitigate the effects of what one report described as the "incremental and involuntary journey towards a Terminator-like reality",[9] referring to the 1984 film in which humans are hunted and killed by robots. Furthermore, while UAS had hitherto been deployed predominantly in the Middle East and South Asia, the Somalia strike represented the opening of a new frontier in the pilotless war campaign, which was already taking place in theatres of conflict, including Afghanistan, Pakistan, and Yemen. By 2019, the number of annual drone strikes in Somalia had peaked at 63, according to data from New America (see Fig. 1). Yet while much of the focus and scholarship on drones has been on such military use, militant non-state groups have been innovating and refining their battlefield tactics, too.

[9] Norton-Taylor, R., "Rise of the drones poses growing dilemma for military", *The Guardian*, 2 April 2012, www.theguardian.com/world/2012/apr/02/rise-of-the-drones-military-dilemma.
[10] New America, "The War in Somalia", *New America*, 2024, www.newamerica.org/future-security/reports/americas-counterterrorism-wars/the-war-in-somalia/.

3 Innovation and Hobbyist Drones

Military-grade MALE drones, including US Predator and Reaper drones, have become the hallmark of America's drone wars. These favoured UAS can fly at high altitudes and loiter over a target—for up to 27 h in the case of the MQ9 Reaper—enabling them to gather intelligence.[11] They are more cost-effective per flying hour than manned aircraft such as F16 fighter jets, but they require substantial military intelligence and physical infrastructure to be launched, piloted, and landed. Such UAS or MALE drones have given militaries significant territorial advantage. They have also bestowed upon global leaders' significant political capital, as drone strikes have been associated with fewer body bags of service men and women. However, the toll of civilian casualties is a highly contested issue. By and large, apart from instances when military-grade drones have been downed and captured by violent non-state actors or donated by proxy forces, as is being observed in Yemen,[12] military-grade drones have remained the preserve of regular armies.

The situation changed in 2016. Nearly 700 km away from where the US 2011 Somalia drone assault had taken place in the port of the city of Kismayo,[13] ISIS militants in Northern Iraq were experimenting with hobbyist or shop-bought drones—the likes of which one might buy for a child as a gift—and using them to lethal effect. In what became known as 'the Battle for Mosul,' the use of such affordable and easily acquired drones against conventional forces for lethal attacks, without doubt, changed operational dynamics by being able to "innovate and surprise its enemies from the air by taking a relatively simple approach".[14]

In October 2016, Kurdish Peshmerga fighters working alongside Iraqi troops in the parched desert around the town of Mosul tried to dislodge ISIS militants. Since 2014, the ISIS fighters had established a firm foothold in the city of Mosul—home to some 1.5 million people according to news reports at the time—and established positions deep in the city. The fighters blended in among the civilian population. Mosul, situated alongside the Tigris River some 400 km north of Baghdad, was a strategic stronghold for ISIS in Northern Iraq and an important ISIS supply route between Iraq and Syria.[15]

[11] Reuters, "What is the MQ-9 Reaper drone that collided with a Russian jet?", *Reuters*, 15 March 2023, www.reuters.com/world/europe/what-is-mq-9-reaper-drone-that-collided-with-russian-jet-2023-03-15/.

[12] Faucon, B., "Houthis Use Sea Drones to Attack Vessels as U.S. Aims to Secure Red Sea", *The Wall Street Journal*, 16 June 2024, www.wsj.com/world/middle-east/houthis-use-sea-drones-to-attack-vessels-as-u-s-aims-to-secure-red-sea-0279cf5e.

[13] Al Jazeera, "US 'extends drone strikes to Somalia'", *Al Jazeera*, 1 July 2011, https://aljazeera.com/news/2011/7/1/us-extends-drone-strikes-to-somalia.

[14] Rassler, D., "The Islamic State and Drones: Supply, Scale, and Future Threats", *Combating Terrorism Center*, 11 July 2018, https://ctc.westpoint.edu/islamic-state-drones-supply-scale-future-threats/.

[15] Hussein, R., "IS Opens New Supply Route Between Syria and Iraq", *VOA*, 9 December 2015, www.voanews.com/a/is-opens-new-supply-route-between-syria-and-iraq/3096047.html.

But as US-backed forces and humanitarian organisations were poised to receive thousands of displaced Iraqis fleeing to shelters or camps as the October 2014 offensive began, they were met with a surprise. The expected civilian exodus did not come. It was initially just a trickle as ISIS fighters occupied key buildings, and civilians found themselves trapped and used by ISIS militiamen as human shields. To the world's surprise, ISIS fighters were able to stave off regular soldiers with the use of, among other things, an arsenal of armed hobbyist drones.[16]

Research, much of it undertaken by Dr Don Rassler from the Combatting Terrorism Center at West Point, has shown that ISIS fighters appear to have used the previous two years to build up a significant stockpile of weapons and, perhaps more significantly, exploit a growing open innovation environment.[17] This was exemplified on 2 October 2016 when a booby-trapped ISIS drone exploded in the city of Mosul, killing two Peshmerga fighters and wounding a French soldier.[18] Although ISIS had been using drones since 2013 for intelligence gathering and surveillance purposes, the offensive at Mosul signalled the group's first forays into applying UAS technology for kinetic use. It demonstrated adaptability and a penchant for technological innovation that Rassler and other researchers like Cronin (2019) and Chávaz and Swed (2021) have used to characterise ISIS and other jihadist groups.[19]

A parallel can perhaps be drawn between the development of hobbyist drone technology and the improvised explosive device (IED), which evolved from a simple tripwire system to remote-controlled detonation using mobile phones. It was a technology widely used in settings such as Iraq and Afghanistan, then adapted for use in the Sahel and other African settings, including Somalia. The IED is often described as the 'poor man's weapon'. As Solène Jomier from the Groupe de recherche et d'information sur la paix et la sécurité (GRIP) commented in an online newspaper article in 2022, "a decade ago, terrorist groups active in the Sahel had only marginal technical expertise in the development of IEDs. It was with the emergence of the Libyan conflict and the fall of the Islamic State in Syria and the Levant (which caused an exodus of fighters to Africa) that this expertise spread to the Sahel".[20] Access to simple components and know-how shared via forums on the internet have helped to diffuse knowledge on simple weapons manufacture.

---

[16] Rassler, "The Islamic State and Drones: Supply, Scale, and Future Threats", 11 July 2018.

[17] Ibid.

[18] Reuters, "Islamic State drone kills two Kurdish fighters, wounds two French soldiers", *Reuters*, 12 October 2016.

[19] Cronin, A.K., "Power to the People: How Open Technological Innovation is Arming Tomorrow's Terrorists", *Oxford University Press*, New York, 2019; Chávez and Swed, "The Proliferation of Drones to Violent Nonstate Actors", 2021.

[20] Dione, I., "IEDs in the Sahel: fearsome war toys for jihadists", *APA News*, 8 November 2022, https://apanews.net/ieds-in-the-sahel-the-fearsome-weapons-for-jihadists/.

Similarly, improvements in battery power and refinements to stabilisation, payload, GPS and ground-to-air control systems[21] have seen the utility of hobbyist drones expand dramatically for malign as well as benign purposes. ISIS was able to exploit this and weaponise hobbyist drones, absorbing these affordable aerial assets into their arsenals. They became an important tactical and strategic weapon, enabling ISIS to project force in a way it had not been able to do before. Indeed, the strategic value of drones in asymmetric warfare enables government forces and irregular fighters to occupy geographical space on the 'battlefield' in a way that was not possible before. For example, in Nigeria, Islamic State West African Province (ISWAP) has reportedly used drones to identify a Nigerian military convoy ahead of an ambush. Likewise, in Mozambique, Islamic State affiliates have used such aerial assets to gather intelligence ahead of attacks.[22] The Nigerian military is also increasingly using drones to fight Boko Haram.[23]

While 2016 would appear to have been a period of experimentation and lesson-learning for ISIS, the group sought to demonstrate a more aggressive posture a year later in 2017. Its fighters launched their first weaponised drone mission over Mosul, targeting an Iraqi outpost and reportedly killing a small group of soldiers.[24]

Over the course of 2016 and 2017, ISIS used drones to assist with precision targeting on the ground and to acquire footage for use in propaganda videos. Deploying drones fitted with cameras served as a recruitment tool by visually depicting acts of destruction using drones. According to Rassler, one video showed "at least a dozen examples of the group releasing munitions on its enemies from the air […] via quadcopter drones".[25]

Furthermore, Cronin (2019) observes that there appeared to have been intense use of UAS by all sides in Iraq and Syria during 2016 and 2017. Citing fellow scholars, she notes that "in the spring of 2017 […] there were between 60 and 100 aerial drone bombing attacks each month across Iraq and Syria," mainly

[21] Hongfei Drone, "Relationship Between Drone Payload and Battery Capacity", 1 August 2023, www.hongfeidrone.com/news/relationship-between-drone-payload-and-battery-capacity/.

[22] Dahiru, A., "How Drones Are Changing The Face Of Terrorism In Africa", *The Human Angle*, 26 February 2024, https://humanglemedia.com/how-drones-are-changing-the-face-of-terrorism-in-africa/.

[23] Runsewe, O., "Nigeria Adopts the 'Drone War' Approach in the Fight Against Boko Haram: Here are other modern warfare options they can take on", *Ventures Africa*, 5 February 2016, https://venturesafrica.com/nigeria-joins-the-drone-wars-in-its-fight-against-boko-haram-here-are-other-weapons-they-could-use/.

[24] Almohammad, A. and Speckhard, A., "ISIS Drones: Evolution, Leadership, Bases, Operations and Logistics", *International Center for the Study of Violent Extremism*, 2017; Warrick, J., "Use of weaponized drones by ISIS spurs terrorism fears", *The Washington Post*, 21 February 2017, www.washingtonpost.com/world/national-security/use-of-weaponized-drones-by-isis-spurs-terrorism-fears/2017/02/21/9d83d51e-f382-11e6-8d72-263470bf0401_story.html.

[25] Rassler, D., "The Islamic State and Drones: Supply, Scale, and Future Threats", *Combating Terrorism Center*, 11 July 2018, https://ctc.westpoint.edu/islamic-state-drones-supply-scale-future-threats/.

in highly urbanised areas.[26] She highlights that ISIS appears to have benefitted from a "global and layered" supply chain that involved "purchases from at least 16 different companies that were based in at least seven different countries".[27] Cronin concludes that the Islamic State's drone programme is an important case study that highlights how the group overcame technical and cost asymmetries and developed a novel weapons system constructed from commercial components that challenged—at least for a time—states' ability to respond.

4 Adaptation and Diffusion Across Africa?

Such intense and rapid escalation in drone use and proliferation in Iraq and Syria—and more recently Russia, Ukraine, Israel and Palestine—has led to growing interest in the use of drone technology by non-state armed groups. But can the African continent expect to see a similar pattern of diffusion and use as described above, or will more conventional weapons such as small arms—which command more attention from regional bodies such as the African Union and its Silence the Guns ambitions[28]—persist?

UAS have, without doubt, been transformative in Africa's economic development and used for overwhelmingly non-malevolent purposes, including for business, law enforcement, weather forecasting, and the delivery of humanitarian assistance and medical supplies.[29] The technology has helped states overcome geographical constraints and cost barriers. For example, in a global market estimated to be worth $43 billion USD[30], non-military use of drones has led to a rapidly expanding and diverse marketplace for manufacturing, supplying and distributing drones. Consequently, malevolent use may be hard to predict, detect or neutralise given the low access costs, dual-use nature, and availability. Yet, in several UN Security Council Resolutions [RES 2370, 2017, 2396 (2017), A/C.1/78/L.56], the lethal threat of drones is increasingly being recognised.[31]

[26] Cronin, "Power to the People: How Open Technological Innovation is Arming Tomorrow's Terrorists", 2019, 218.

[27] Rassler, D., "The Islamic State and Drones: Supply, Scale, and Future Threats", *Combating Terrorism Center*, 11 July 2018, iv.

[28] African Union, "Silencing the Guns: Creating Conducive Conditions for Africa's Development", n.d., https://au.int/en/flagships/silencing-guns-2020.

[29] Do4Africa, "The drone industry in Africa", n.d., www.do4africa.org/en/the-drone-industry-in-africa/.

[30] Drone Industry Insights, "The South African Drone Market Report 2019–2024", *DRONEII*, 2024, https://droneii.com/product/south-african-drone-market-report.

[31] United Nations, "A/C.1/78/L.56: First Committee Approves New Resolution on Lethal Autonomous Weapons, as Speaker Warns 'An Algorithm Must Not Be in Full Control of Decisions Involving Killing'", Meetings Coverage and Press Releases, 1 November 2023, https://press.un.org/en/2023/gadis3731.doc.htm.

In its Resolution 2370 (2017), the UN Security Council strongly condemns the continued flow of weapons, including small arms and light weapons (SALW), military equipment, UAS and their components, and IED components to and between ISIL, al-Qaeda, their affiliates and associated groups, illegal armed groups, and criminals. It encourages states to prevent and disrupt their procurement networks for such weapons, systems and components. The threat of terrorist use of UAS also requires states to enhance measures to counter terrorist targeting of critical infrastructure and vulnerable and soft targets. In its Resolution 2341 (2017), the UN Security Council encourages states to make concerted and coordinated efforts to raise awareness and expand knowledge of challenges posed by terrorist attacks to be better prepared for such attacks. In its Resolution 2396 (2017), the Council underscores the need for states to develop, review, or amend national risk and threat assessments to enhance their preparedness for potential attacks against soft targets. In the same resolution, the Council highlights the need to establish and strengthen national, regional, and international partnerships with public and private stakeholders in this context. Guiding Principles 15, 16 and 17 of the Addendum to the guiding principles on foreign terrorist fighters (2018) are especially relevant to states' efforts to prevent and counter UAS attacks against critical infrastructure, vulnerable targets, soft targets, and tourism, as well as to preventing and combating the illicit trafficking of SALW.

Additionally, the International Civil Aviation Organization (ICAO) has developed the UAS Toolkit to assist states in developing effective UAS operational guidance and safe domestic operations in this area. Security Council Resolution 2309 (2016) on terrorist threats to civil aviation should also be considered in this regard (see Fig. 2). States face persistent challenges with respect to sharing key information on recent incidents, attack methods, and landside gaps at airport facilities.

A 2024 study by Bárbara Morais Figuerido for the UN Institute for Disarmament Research (UNIDIR) has sought to capture the current state of drone use by violent non-state armed groups in Africa. Although dependent on publicly available information from within the UN system and think tanks, it is an important study which sheds light on trends. The study concludes that "as unarmed and armed UAS become more accessible and advanced, and as these systems are deployed by an increasingly diverse set of actors across Africa, the risks posed by NSAGs' use of UAS to national and regional security could become more acute and should be monitored closely".[32]

The African continent may not have experienced the rapid escalation of drone use by armed non-state groups on the scale observed in Iraq and Syria yet. Still, a dramatic rise in the deployment of drones in various may serve as an inspiration in the future and merit further research both globally and in the African context.

[32] Figueiredo, "The Use of Uncrewed Aerial Systems by Non-State Armed Groups: Exploring Trends in Africa", *UNIDIR*, 30 January 2024.

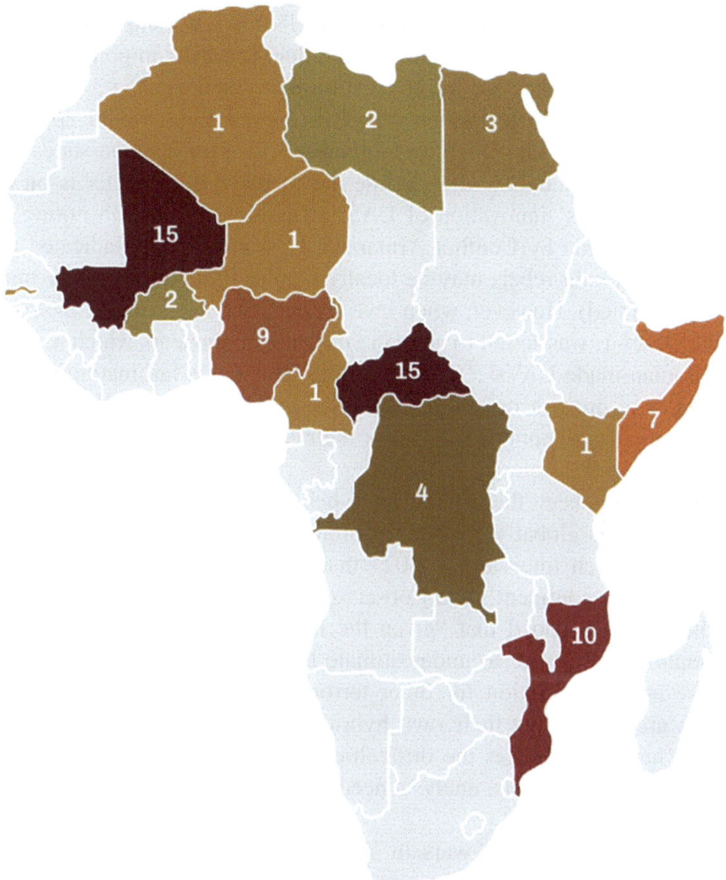

Fig. 2 Recorded incidents of use of UASs by NSAGs in Africa, by State, January 2018–June 2024. *Source* UNIDIR[33]

At the time of writing, an escalation in Houthi rebel attacks in the Red Sea[34] would appear to indicate that drone warfare by non-state armed groups is becoming a persistent feature in Africa-focused conflicts both on land and at sea. Denys Reva from the Institute for Security Studies, South Africa, warns that "given existing maritime security challenges and inadequate border protection," African countries

[33] Figueiredo, B.M., "The Use of Uncrewed Aerial Systems by Non-State Armed Groups: Exploring Trends in Africa", *UNIDIR*, 30 January 2024, 13, https://unidir.org/publication/the-use-of-uncrewed-aerial-systems-by-non-state-armed-groups-exploring-trends-in-africa/.

[34] Voice of America (VOA) News, "US forces destroy six Houthi drones in Red Sea", *VOA News*, 21 June 2024, www.voanews.com/a/us-forces-destroy-six-houthi-drones-in-red-sea/7664960.html.

may be insufficiently prepared.[35] In the abovementioned incidents of Houthi rebel attacks, where the target appears to have been US Navy warships, the underwater unmanned vehicle used would appear to have had several components similar to those manufactured in Iran, according to US military sources.[36]

The acquisition of the technology used in these attacks would appear to be the result of both local adaptation and alliances with proxy nation-states who act as sponsors in the Yemeni conflict. While the focus of this chapter is on Africa-based non-state actors' innovation of UAS technology, it is worth noting that an extensive investigation by Conflict Armament Research (CAR) indicated that the drones used by Houthi rebels may be locally manufactured (or partially manufactured and assembled). However, when investigated by CAR experts, one particular model, the Qasef-1, was found to contain "internal components which match those found in Iranian-made UAVs".[37] Although this provides a fascinating insight into the geopolitics of drone acquisition, adaptation and proliferation, the focus of this chapter is on the use of hobbyist drones or their components rather than on weapon supplies from friendly states.

Rassler, who has been the driving force behind research into UAS use by violent non-state actors globally, told the Institute for Security Studies, South Africa, during a pre-research interview in 2021 that it may only be "a matter of time" before the African continent would observe similar tactics to those witnessed in the Middle East. He added that "given the Islamic State's penchant for innovation, it would be a mistake to underestimate the organisation and how the group could serve as an inspiration for other terrorists and/or nation-states and proxy groups that are developing their own hybrid warfare strategies. However, in his 2016 paper, he acknowledges the difficulties of making predictions and cautions that to monitor future threats, analysts need to be "part terrorist, part technologist and part industry expert".[38]

Notwithstanding the proxy wars in Yemen and Libya—where Egypt, Russia, France, Qatar, Turkey, UAE, Italy, and the USA have all played a part—Africa has not seen a surge in drone use by violent non-state actors on the scale seen in the Middle East. This has stimulated research into the factors determining acquisition, adaptation, and proliferation. Network affiliation would appear to be a factor, according to UNIDIR research,[39] such that Africa-based armed groups with links

[35] Reva, D., "Abandon ship: uncrewed vessels threaten Africa's maritime security", *Institute for Security Studies*, 4 April 2024, https://issafrica.org/iss-today/abandon-ship-uncrewed-vessels-threaten-africa-s-maritime-security.

[36] US Central Command, "CENTCOM Intercepts Iranian Weapons Shipment Intended for Houthis", 15 February 2024, www.centcom.mil/MEDIA/PRESS-RELEASES/Press-Release-View/Article/3677794/centcom-intercepts-iranian-weapons-shipment-intended-for-houthis/.

[37] Conflict Armament Research Ltd., "Evolution of UAVs employed by Houthi forces in Yemen", *Storymaps*, 2020.

[38] Rassler, "Remotely Piloted Innovation: Terrorism, Drones and Supportive Technology", 2016, 64.

[39] Figueiredo, "The Use of Uncrewed Aerial Systems by Non-State Armed Groups: Exploring Trends in Africa", *UNIDIR*, 30 January 2024.

to al-Qaeda or ISIS appear more likely to use drone technology. However, more research is needed on the topic.

Another factor which may cloud research on narrow classifications such as hobbyist drones and violent non-state actors in Africa is the blurring of the distinction between drones used by militaries and drones used by armed groups. For instance, regular armed forces have at times resorted to hobbyist drones at an operational and tactical level, including in settings such as Ethiopia.[40] At the same time, non-state groups have acquired drones through sponsored support by proxies or through capturing enemy assets.

5 Incidents of the Use of Drones by Violent Non-state Actors in Africa

UAS manufacturing, adaptation and usage have rapidly expanded across Africa, a phenomenon captured in a key report, *Remote Horizons*, by the Dutch organisation PAX for Peace. Focused largely on military-grade drones, *Remote Horizons* notes that since 2011, "drone operations have been carried out in at least 20 states, of which armed drones have been used in 9 states across Africa.[41] An earlier report by PAX for Peace, Violent Skies also notes the growth in the commercial drone sector 'has contributed to the proliferation of non-state drone warfare".[42]

The secrecy surrounding the use of UAS technology by militaries across Africa, along with limited reporting of drone deployments, have resulted in what PAX terms a "fog of drone war in Africa".[43] This, in turn, has led to a lack of oversight on using this technology and a lack of preparedness in identifying risks and building countermeasures.

The risks may not be limited to conflict settings but also exposures associated with critical infrastructure technology (ports, airports, or communications centres, for example), which may be targeted by groups outside a theatre of conflict. One counter-drone expert told this author in February 2024 that he fears the first incident in Africa "when a UAS with explosives attached to it flies into a bank or another piece of critical infrastructure."[44] This could pose a bigger threat than such an incident on a battlefield. In some settings, such as South Africa, there is a burgeoning marketplace for counter-drone technology to protect private assets

[40] Atherton, K.D., "Mass-Market Military Drones Have Changed the Way Wars are Fought", *MIT Technology Review*, 30 January 2023, www.technologyreview.com/2023/01/30/1067348/mass-market-military-drones-have-changed-the-way-wars-are-fought/.

[41] Kurpershoek, R., Valdez, A.M. and Zwijnenburg, W., "Remote Horizons: Expanding use and proliferation of military drones in Africa", *PAX for Peace*, February 2021.

[42] Zwijnenburg, W. and Jansen, A., "Violent Skies: How lethal drone technology is shaping contemporary warfare", *PAX for Peace*, 29 November 2020, https://paxforpeace.nl/wp-content/uploads/sites/2/import/import/paxviolentskies_0.pdf.

[43] Ibid.

[44] Ibid.

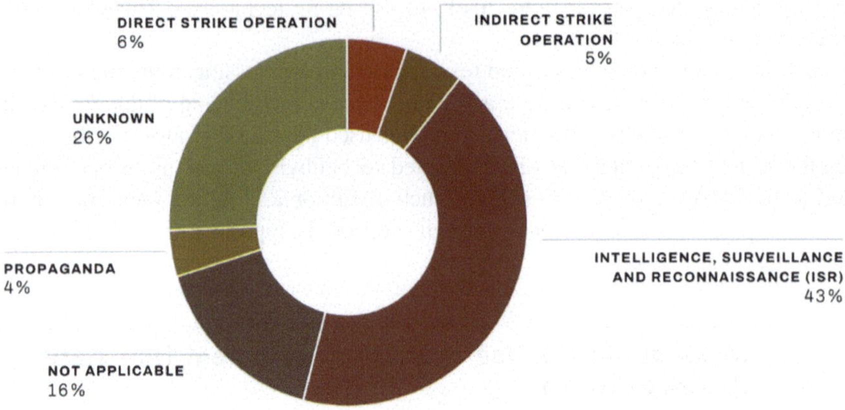

Fig. 3 Types of use of UAS by NSAGs in Africa, January 2018–June 2023. *Source* UNIDIR[46]

such as mines, energy plants and other critical infrastructure. This has possibly been inspired by the experience of explosives being used in cash-in-transit heists led by transnational criminal networks[45] and concerns about the nexus between these and terrorist groups. However, notwithstanding the limited inroads that have been made into developing a growing situational awareness about malign uses of drone technology, an absence of leadership, political will, and policy competition across the African continent limits African states' ability to engage publicly on the potential threat. Consequently, much of the reporting on drone deployments (both by state and violent non-state actors) comes from media reports, industry sources or eyewitness accounts.

Based on monitoring from sources such as Vision of Humanity's Global Terrorism Index and PAX for Peace, the overwhelming use of hobbyist UAS by violent non-state actors is for intelligence, surveillance, and reconnaissance (ISR) purposes rather than kinetic use. UNIDIR has classified use in the graphic (Fig. 3)

6 Target Acquisition and Direct Strikes

In 2020, the Somalia-based al-Qaeda affiliate al-Shabaab, used drones to coordinate attacks on a military base in neighbouring Kenya at Manda Bay. The attack, which killed three service personnel and civilians, also destroyed six US aircraft

[45] Schmidt, M., "Explosives-based crime costs the economy tens of billions of rand", *Business Day*, 23 October 2023, www.businesslive.co.za/bd/national/2023-10-23-explosives-based-crime-costs-the-economy-tens-of-billions-of-rand/.

[46] Figueiredo, "The Use of Uncrewed Aerial Systems by Non-State Armed Groups: Exploring Trends in Africa", *UNIDIR*, 30 January 2024, 12.

and several vehicles, according to a US investigation which followed.[47] The group subsequently used footage acquired from the attack for propaganda purposes[48], enabling it to project al-Shabaab's presence far beyond the borders of Kenya and Somalia. After this attack, the UN Security Council reported that member states had seized manuals for the use of drone technology in targeted attacks. Such manuals are now widely available on the internet.[49]

Armed groups across Africa have also used drones for target acquisition and precision targeting. In Nigeria in 2022, Islamic State West Africa Province (ISWAP) reportedly used surveillance drones in positions above Nigerian state forces shortly before its fighters launched an attack.[50] Furthermore, Malik Samuel (Institute for Security Studies researcher) says research shows that ISWAP "is already trialling delivery drones to carry improvised explosive devices. This includes assessing the weight that can be carried, how far they can go and how long they can stay in the air".[51]

In Mozambique's Cabo Delgardo Province, hobbyist drones have been used by militants for precision targeting both on land and at sea. In 2020 and 2021, during the peak of counter-insurgency,[52] operations in Northern Mozambique against the militant group Ahlu-Sunnah wal Jama'a (ASWJ) were supported by the Islamic State in Mozambique.[53] Interior Minister Amande Miquidade reported that insurgents from ASWJ had targeted the strategic town of Palma to "improve attack precision". Other individuals operating in the region, including retired Col David Peddle, a former member of the South African National Defence Force, in an interview I conducted with the Institute for Security Studies, South Africa in 2020, corroborated this. Furthermore, Peddle warned that 'drone swarms' or clusters of offensive drones across Africa could become the next iteration of the application of the technology. Although, at the time of writing, the tactical use of drone swarms has not been observed on the continent.

---

[47] Lopez, C.T., "At Manda Bay, Investigation Finds No Single Point of Failure, But Many Recommendations for Improvement", US Department of Defense, 10 March 2022, www.defense.gov/News/News-Stories/Article/Article/2963234/at-manda-bay-investigation-finds-no-single-point-of-failure-but-many-recommenda/.

[48] Dass, R., "The Evolving Threat From Terrorist Drones in Africa", *Lawfare*, 1 May 2023, www.lawfaremedia.org/article/the-evolving-threat-from-terrorist-drones-in-africa.

[49] Chávez and Swed, "The Proliferation of Drones to Violent Nonstate Actors", 2021.

[50] Obiezu, T., "Regional Security Analysts Say Africa at Risk of Drone Terrorism", *VOA News*, 22 November 2023, www.voanews.com/a/regional-security-analysts-say-africa-at-risk-of-drone-terrorism-/7366009.html.

[51] Samuel, M., "ISWAP's use of tech could prolong Lake Chad Basin violence", *Institute for Security Studies*, 13 April 2023, https://issafrica.org/iss-today/iswaps-use-of-tech-could-prolong-lake-chad-basin-violence.

[52] Louw-Vaudran, L., "The many roots of Mozambique's deadly insurgency", *Institute for Security Studies*, 8 September 2022, https://issafrica.org/iss-today/the-many-roots-of-mozambiques-deadly-insurgency.

[53] International Crisis Group, "Mozambique", n.d., www.crisisgroup.org/africa/east-and-southern-africa/mozambique.

The fear is not simply of the use of UAS on the battlefield but also in civilian-focused attacks by armed groups on critical infrastructure such as ports and airports. A very illustrative example of this was the Houthi drone attack on an oil port in Saudi Arabia in 2023.[54] With the growing offshore oil and gas industry on the east coast (Mozambique) and west coast (Namibia, Angola, Nigeria, etc.) of Africa, countering drone attacks must be planned for by government.

7 Propaganda

The recent use of UAS to acquire visual media useful for propaganda campaigns has been a tactic used by armed groups. Some examples include the Islamic State in West Africa Province (ISWAP) in Nigeria and the Lake Chad Basin between January and April 2022 (Dass & Singh, Islamic State in West Africa Province Video Signals the Group's Grand Strategy for the Future, 2022). In the latter part of 2022, Islamic State Central Africa Province (ISCAP) released a propaganda video that was filmed with the use of a drone. A Mali-based al-Qaeda affiliate, Jama'at Nusrat al-Islam wa al Muslemeen (JNIM), has collected filmed content from commercial drones to depict its training camps in an apparent recruitment drive.[55]

As hobbyist drone acquisition proliferates and the platforms for disseminating propaganda material—including social media platforms and messaging platforms such as Telegram—expand, one can assume the potential to use drones to collect visual imagery will increase. At the time of writing, the organisation Tech Against Terrorism is collecting data on the use of terrorist propaganda online as well as the use of generative AI (synthetic media) to amplify their narratives.[56] Some of the raw footage will without doubt be collected via drones, with some of it exhibited on websites which the insurgent groups control and own.[57]

8 Intelligence, Surveillance and Reconnaissance

Across all the conflict zones cited above including the Sahel, Lake Chad Basin, the Horn of Africa and the Democratic Republic of Congo (DRC), the predominant use of drones by armed groups has been for surveillance purposes to gain situational awareness or intimidate the civilian population. For example, in the DRC,

[54] Al Jazeera, "Saudi Aramco's Jeddah oil depot hit by Houthi attack", *Al Jazeera*, 26 March 2022, www.aljazeera.com/news/2022/3/25/saudi-aramco-jeddah-storage-facility-hit-by-attack.

[55] Balkan, S., "A Global Battlefield? Rising drone capabilities of non-state armed groups and terrorist organizations", *SETA*, Ankara, 2019.

[56] Tech Against Terrorism, "Early Terrorist Adoption of Generative AI", *Tech Against Terrorism*, 8 November 2023, https://techagainstterrorism.org/news/early-terrorist-adoption-of-generative-ai.

[57] Tech Against Terrorism, "Terrorist-Operated Websites", 2022, https://techagainstterrorism.org/terrorist-operated-websites.

the UN reports that the sanctioned group, the Allied Democratic Forces (ADF), has used drones for surveillance purposes since 2021. According to testimonies from former combatants and captives, ADF used UAS to monitor their camp's surroundings when foodstuffs, medicine, weaponry or IED components were being brought by ADF collaborators. They claimed meetings had been held in the South Kivu province of the DRC between senior ADF commanders and Ahl al-Sunna wal-Jama'a (ASWJ) operatives, an ISIS-linked group which had operations in Mozambique. They also "highlight attempts by Islamic State operatives based in South Africa to transfer commercially available UAS to the ADF".[58] This example underscores why some scholarship has focused on the transfer of knowledge and technology between armed groups, with Chávez noting that "network affiliations are the strongest predictors of adopting a drone programme".[59]

9 Proliferation Prospects Across Africa

Given the porous nature of borders across the African continent and existing weapons smuggling routes, the prospect of UAS proliferation across the rest of Africa is likely. Not only are hobbyist, dual-use technologies easy to purchase through traditional retail outlets or online, but the wholesale movement of large batches of stock or components of more sophisticated drones by transnational criminal networks with links to jihadist groups may exploit existing loopholes which facilitate conventional arms proliferation.

Dass, citing an October 2022 US investigation revealing weapons components and IEDs being supplied from Yemen to al-Shabaab militants in Somalia, argues that although there is no evidence yet of drones having been smuggled, "the likelihood of similar networks transferring drone technology from Yemen into Africa is possible given the prevalence of drone technology that is already present in Yemen and used by the Houthis".[60]

The proliferation of drone technology for benign and benevolent use, such as for humanitarian purposes, enables clandestine action by threat actors. A Chinese-made DJI drone can be purchased to deliver lifesaving medicines or repurposed to drop a small grenade for as little as a few hundred US dollars. The dual-use nature, low cost, and accessibility of the technology enables it to evade the limitations of the existing regulations. Regulations are limited and broadly apply to the larger military grade drones which are bound by international humanitarian law, while civilian regulation is under the auspices of the International Civil Aviation

[58] Group of Experts on the Democratic Republic of the Congo, "Letter dated 13 June 2023 from the Group of Experts on the Democratic Republic of the Congo addressed to the President of the Security Council", *ECOI*, 13 June 2023, www.ecoi.net/en/file/local/2094036/N2312380.pdf.

[59] Allen, K., "Drones as weapons: Africa needs better data to anticipate risk", *Institute for Security Studies*, 17 November 2023, https://issafrica.org/iss-today/drones-as-weapons-africa-needs-better-data-to-anticipate-risk.

[60] Dass, "The Evolving Threat From Terrorist Drones in Africa", *Lawfare*, 1 May 2023.

Organisation (ICAO), which has focused broadly on safety, privacy and data issues. While the topic of regulation is beyond the scope of this chapter, the increased use of hobbyist or commercial UAS by non-state armed groups has sparked a lively debate with advisors to the International Committee of the Red Cross, urging technologists and policy makers to work together to come up with solutions. One such proposed solution is ID markers to identify UAS used for humanitarian as well as potentially malign purposes.[61]

The greater proliferation of UAS by non-state armed groups for malign purposes poses an existential threat to African militaries. As a senior Senegal Air Force chief of staff conceded at a major gathering in the country in 2023, the rapid evolution of drone technology represents a "paradigm shift" and is revolutionising the balance of power.[62] It challenges the Webbian philosophy that the state (regular armies) has the monopoly on the use of violence in war by effectively levelling the playing field on land, in the air and at sea.

In contemporary conflicts beyond Africa, the modification of hobbyist or commercial off-the-shelf drones is increasingly becoming a feature in combat settings. Armed groups in Africa are doubtlessly watching and learning as new iterations of UAS technologies emerge.

Chávez and Swed note that the appeal of hobbyist drones to violent non-state actors is their versatility. They can "use a single platform to perform multiple tasks—surveillance, target acquisition, propaganda dissemination, and even to drop munitions at a distance".[63] They also narrow the gap between conventional military capabilities and those of irregular forces as stated above. In other words, they represent an asymmetric force multiplier. Regular militaries have long enjoyed air superiority against non-state actors, but the advent of drone technology has arguably eroded this advantage. It also now enables violent non-state actors to project power into the aerial domain. They have forced militaries to expend energy and time on countermeasures. By 2021, the scholars noted that some 57 violent non-state actors are using drones globally.[64]

10 Drone Technology and the Future

While technological advances in hobbyist UAS are enabling enhancements to their flying and loitering capacity, the advent of other technologies such as Artificial Intelligence may assist in shaping their future use. Alysa Simms highlights a

[61] Greenwood, F., "Consumer Drones in Conflict: where do they fit into IHL?", *International Committee of the Red Cross*, 15 March 2022, https://blogs.icrc.org/law-and-policy/2022/03/15/consumer-drones-conflict-ihl/.

[62] defenceWeb, "Second Africa Air Force Forum kicks off in Senegal", *defenceWeb*, 26 October 2023, www.defenceweb.co.za/aerospace/aerospace-aerospace/second-africa-air-force-forum-kicks-off-in-senegal/.

[63] Chávez and Swed, "The Proliferation of Drones to Violent Nonstate Actors", 2021.

[64] Ibid.

model of UAS that the Chinese company DJI developed as one of the biggest consumerist drone manufacturers in the world. She describes how this model can enable the user to monitor up to 16 moving targets simultaneously and upload data to the internet in real-time.[65] Furthermore, as artificial intelligence (AI) capabilities expand, the ability of drones to seek out targets independent of human direction and oversight increases.

One South Africa-based expert interviewed for this chapter in 2024 warned that self-built kamikaze drone swarms could post an enormous threat in future. In his words they could "be pre-programmed, sent on a path, and switched off, making them hard to track by sensors", Sounding a warning to leaders and technologists alike he added "this is a potential iteration of the technology that policymakers need to be aware of, and as AI innovation advances, so do the potential threats."

Furthermore, as an arms race in technology emerges, Zachary Kallenborn, a part-time fellow at the Center for Strategic and International Studies' Strategic Technology Program (CSIS), highlights the issue of scalability. Kallenborn is of the opinion that software advances could help extremists counter governments' military-grade technology by using dozens or hundreds of commercially available drones to create swarm.[66] The problem is compounded further by 3D printing becoming more widely available in Africa; and with it, the possibility of printing UAS and their components.

As well as exploiting new technology, armed groups using UAS in Africa may also consider new targets beyond conflict zones. For example, aerial assaults on critical infrastructure similar to those reported in other settings including the United States[67] may pose a potential threat. Already, cyber-attacks on pieces of critical infrastructure such as the Transnet port in South Africa[68] have demonstrated how threat actors can disrupt essential state functions and supply chains that extend beyond a state's borders and across the region. A UAS laden with explosives flown into port infrastructure could similarly inflict damage to state assets and public perceptions of safety, as well as the capacity of the state to protect its citizens. Such attacks could furthermore serve to provoke a state response to a perceived adversary, threatening an escalation in pre-existing hostilities or political tensions.

[65] Simms, A., "The Rising Drone Threat from Terrorists", *Georgetown Journal of International Affairs* 19, 2018.

[66] Africa Defense Forum, "The dizzying possibilities of drones in Africa", *defenceWeb*, 20 October 2023, www.defenceweb.co.za/aerospace/aerospace-aerospace/the-dizzying-possibilities-of-drones-in-africa/.

[67] Starr, J., "Addressing the Threat of Hostile Drones to Critical Infrastructure", *C4ISRNet*, 1 March 2023, www.c4isrnet.com/opinion/2023/03/01/addressing-the-threat-of-hostile-drones-to-critical-infrastructure/.

[68] Shabalala, Z. and Heiberg, T., "Cyber attack disrupts major South African port operations", *Reuters*, 22 July 2021, www.reuters.com/world/africa/exclusive-south-africas-transnet-hit-by-cyber-attack-sources-2021-07-22/.

11 Conclusion

This chapter has served to locate the use of hobbyist UAS by non-state armed groups in Africa in a global context and discuss the likelihood of further proliferation based on what has been observed in other settings.

Even if the African continent has not experienced the degree of proliferation as observed in other conflicts zones such as the Middle East and in the current Russia-Ukraine and Israel-Palestine conflicts, the utility and attractiveness of UAS from the point of view of a non-state armed group is significant. Indeed, it is the disruptive nature of this innovation which gives unprecedented capabilities to groups and individuals who would otherwise not have any aerial surveillance and attack capability. Ultimately this affects the relationship of adversaries in hybrid war settings. It also adds to the costs of regular militaries who must bolster their air defences to account for the fact that UAS diminish asymmetry advantage. Furthermore, the technology, when weaponised, offers VNSAs the "capacity to use lethal force from the air at very low altitudes that are typically unprotected by military air defenses".[69]

While the data shows that the number of kinetic attacks on the African continent using UAS is relatively low, the potential to exploit online information sources and network affiliations and find greater utility in drones, and possibly new use, could see their deployment escalate in Africa. For example, armed groups may scale up their use of drones to execute surveillance and reconnaissance tasks, but also to gather material for propaganda purposes or to recruit more individuals into their networks. AI enhancements, both in terms of how visual material is gathered by UAS as well as how that material is disseminated, may make this the most attractive use case in the future in Africa.

This chapter has sought to provide context to the malign use of hobbyist off-the-shelf UAS by armed groups in Africa as part of a rapidly emerging global drone ecosystem. In much of that ecosystem UAS use is benign. Indeed, the utility of UAS technology to advance economic development and social welfare across Africa has been well documented. However, regional and continental bodies such as the African Union must not be blind to the threat of hobbyist UAS being deployed for malevolent ends, and consider protective measures as part of a wider global discussion, in order to harness the benefits.

Karen Allen is an award-winning former BBC Foreign Correspondent with nearly three decades of industry experience in conflict, humanitarian response, politics, economics, and general news reporting. Allen was posted to Iraq, Kenya, South Africa and Afghanistan during her career as a staff correspondent with the BBC and used these bases to report extensively across Africa and the Middle East for the BBC's TV, Radio and Digital networks. Allen has a Master's degree in International Relations and Contemporary war and is a visiting fellow in the Department of War

[69] Cronin, "Power to the People: How Open Technological Innovation is Arming Tomorrow's Terrorists", 2019.

Studies at King's College London. She has undertaken studies at Harvard University and maintains a close network with several African universities. She undertakes consultancy work for various clients across Africa, Europe, and the United States.

Researching Drones in African Battlespaces: Broader Implications and Policy Recommendations

Wayne Stephen Coetzee and Dries Putter

Abstract

This chapter examines the complex implications of drone technology deployment in African battlespaces, underscoring its dual-use character that simultaneously facilitates humanitarian efforts and exacerbates security threats posed by state and non-state actors. It emphasizes the critical role drones play in reshaping geopolitical dynamics, creating new forms of asymmetry, and generating distinct ethical, legal, and socio-political challenges specific to Africa. Moving beyond conventional military analyses predominantly centred on Western contexts, this contribution highlights the unique African operational realities, stressing the urgent need for tailored ethical guidelines and comprehensive regulatory frameworks. The chapter proposes actionable policy recommendations, including developing indigenous drone capabilities, establishing robust regulatory environments, promoting ethical accountability in drone warfare, fostering regional cooperation, reducing dependency through diversified partnerships, enhancing cybersecurity and counter-drone capabilities, and expanding civilian drone applications. Ultimately, this chapter aims to inform policymakers and stakeholders, advocating for African-led international dialogue and sustainable governance of drone technologies within the continent's diverse security landscape.

W. S. Coetzee (✉)
School of Business, Economics and IT, University West, Trollhättan, Sweden
e-mail: wayne-stephen.coetzee@hv.se

D. Putter
Faculty of Military Science, Stellenbosch University, Saldanha, South Africa
e-mail: putter@sun.ac.za

W. S. Coetzee and D. Putter (eds.), *Drones in the African Battlespaces*, Southern Space Studies, https://doi.org/10.1007/978-3-031-87147-4_9

1 Overview

The dual-use nature of drones—facilitating both peacekeeping and humanitarian missions while enabling state and non-state actors to utilise them as instruments for various combat operations establishes these tools as a double-edged sword within African battlespaces. Throughout this volume, it has been asserted that the accessibility and affordability of drone technology enable African states and non-state actors to engage in various operations with limited resources, leading to the emergence of new asymmetries within African battlespaces and the conflicts therein. However, this accessibility has led to unintended consequences, including the emergence of drone-enabled terrorism and insurgency tactics. Terrorist organisations, such as Boko Haram and al-Shabaab, among others, are procuring drones and modifying them for surveillance in support of propaganda and even kinetic actions. This presents novel challenges for counterterrorism initiatives in a region already burdened by asymmetric threats.

Upon collective examination, the chapters within this volume endeavour to provide a comprehensive and contextually nuanced understanding of the utilisation of drones in African battlespaces. They intend to surpass conventional perceptions that have often focused solely on the military applications of drones, frequently neglecting the unique geopolitical and operational realities that exist in Africa. Consequently, this work contributes to the field by offering insights that are specific to Africa, contrasting significantly with the more extensive, often Western-centric corpus of research on drone warfare. It is evident that drones in African battlespaces encompass more than merely a tactical evolution; they represent a substantial shift in how states, international powers, and non-state actors exert influence within a globalised and technology-driven environment. While drones may confer strategic advantages in certain instances, they simultaneously introduce intricate ethical, legal, and socio-political challenges, particularly given that many African states remain predominantly reliant on foreign suppliers for advanced drone technologies.

A prevalent theme that emerges throughout the chapters in this volume is the imperative for ethical standards and regulatory frameworks designed explicitly for African contexts. These recommendations emphasise the importance of sustaining public trust, respecting state sovereignty, and preventing violations of human rights—representing a shift from some prior research that concentrated exclusively on drones' efficacy (kinetic or non-kinetic). The insights articulated herein outline a strategic pathway for action: African states are encouraged to address dependencies, create resilient regulatory frameworks, and participate in the global discourse on the ethical application of drones. By adopting a proactive approach to these issues, African countries possess the potential to position drones not merely as tools for defence and security but also as instruments of diplomacy, thus advancing security objectives while contributing to a more equitable and balanced regional and global security architecture.

In conclusion, we propose seven policy-relevant recommendations for African governments, drawn from the insights of this volume's chapters and related research. These recommendations include practical steps and expected outcomes.

1.1 Invest in Indigenous Drone Capabilities and Workforce Development

African governments should establish national programmes to fund research and development in unmanned systems and associated technologies. Collaborations with universities and private technology firms could facilitate the establishment of drone innovation hubs, thereby fostering local expertise in the design, manufacturing, maintenance, and deployment of drones. Developing indigenous drone capabilities diminishes reliance on foreign entities, ensures enhanced control over security operations and security of value- and supply chains, and generates employment opportunities, contributing to economic development. Some countries have already begun efforts to explore local drone production. Expanding this paradigm throughout the continent could foster sustainable defence technology industries.

During the COVID-19 pandemic, Uganda, supported by the UN Capital Development Fund (UNCDF) and the Swedish International Development Cooperation Agency (SIDA), collaborated with the Makerere University Infectious Diseases Institute (Uganda) to launch a pilot program that uses medical drone technology in the West Nile districts of Moyo and Adjumani.[1] The project aimed to provide surveillance capabilities for pre-emptive reporting on emerging threats and to offer rapid response capabilities during the COVID-19 pandemic.[2]

Similar cases of drone technology use can be found in Malawi. Malawi is leading the way in utilising drones to deliver medical supplies and conduct medical research in remote areas of the country. Malawi University of Science and Technology (MUST) is collaborating with Virginia Polytechnic Institute and State University (Virginia Tech) through the African Drone and Data Academy—a program sponsored by the United Nations International Children's Emergency Fund (UNICEF). The program facilitates innovation and pilot training within the context of unmanned systems and technology. Highlighting Africa's resilience, reports indicated that during the project's initial training and educational stages, "local students—who had never seen a drone before—built their own drone within a couple

[1] UNCDF, "Leveraging Drone Technology in the COVID-19 Rapid Response in Rural Areas of Uganda", United Nations Capital Development Fund, 2 November 2021, www.uncdf.org/article/7275/leveraging-drone-technology-in-the-covid-19-rapid-response-in-rural-areas-of-uganda.

[2] Ibid.

of days. And it was one of these drones that completed a fully autonomous drug delivery flight over some 19 km”.[3]

Another instance of leapfrogging development cycles can be seen in Rwanda. The Rwandan Information Society Authority (RISA) is working to create a world-class Drone Operation Center (DOC) at Huye Airport (Rwanda). The DOC acknowledges that drones and related technologies will result in “innovation, education, and economic development” and seeks to establish Rwanda “as a regional leader in drone technology”.[4] Key partners in this project include Poland, France, Switzerland, the United Arab Emirates, the US, Japan, and others, providing technology and training services, along with the Malawi-based African Drone and Data Academy.[5]

South Africa is also no stranger to combining drones with university education and innovation programmes. Since 2022, Nelson Mandela University has been at the forefront of this initiative in South Africa, becoming the “first university in South Africa to legally operate drones”.[6] Karl du Preez (Director and Principal Lecturer of the University’s Advanced Mechatronic Technology Centre (AMTC), envisions that “[f]isheries, illegal poaching, whale populations, forestry and fire control” are among the domains that will benefit immensely from drone technology.[7]

Drone pilot training is currently receiving significant attention in several African countries. Among these are Botswana, led by the Botswana International University of Science and Technology and Drones for Africa. Botswana recognises that the “drone industry presents significant opportunities for research, growth and innovation in various sectors of the economy including agriculture, mining, infrastructure development and medicine, just to mention but a few”.[8]

Since 2023, the list of drone training centres across Africa has expanded rapidly with several new business ventures in addition to those already mentioned in South Africa, Zimbabwe, and Namibia.[9] While refraining from cataloguing every ongoing initiative, these developments and efforts by businesses, international partners, and local academic institutions exemplify the growth of the drone market and the

---

[3] UNICEF, “The Development of UNICEF’s Drones and Data for good initiative”, The African Drone and Data Academy, Malawi, 2023, www.unicef.org/innovation/AfricanDroneAcademy and www.unicef.org/malawi/media-coverage-african-drone-and-data-academy.

[4] RISA, “Drone Operation Center (DOC)”, Rwanda Information Society Authority, 2024, www.risa.gov.rw/projects/drone-operation-center-doc.

[5] Ibid.

[6] Nelson Mandela University, “The first university to legally operate drones for a better world”, 10 October 2022, https://news.mandela.ac.za/News/The-first-university-to-legally-operate-drones-for.

[7] Ibid.

[8] Vice Chancellor, Professor Otlogetswe Totolo statement. (BIUST, “BIUST and drones for Africa signed an MOU”, 9 June 2023, www.biust.ac.bw/2023/06/09/biust-and-drones-for-africa-signed-an-mou/).

[9] Drones.R.Africa, “Drone training schools in Africa”, *Drones.R.Africa*, 7 May 2023, https://dronenews.africa/drone-training-schools-in-africa/.

development of human capital within the industry in Africa. Nonetheless, all these instances indicate that, in the absence of a robust national and regional regulatory framework, the environments occupied by drones may present significant risks to security and stability on the continent.

1.2 Implement Comprehensive Regulatory Frameworks for Drone Usage

Meyer and Haafizah state that drones, "technically known as remotely piloted aircraft systems (RPAS) or unmanned aircraft (UA), are transforming various industries, including mining, agriculture, construction, environmental monitoring, emergency response, real estate and infrastructure inspection. In light of this, it is necessary for both private and commercial operators to understand the laws governing drone usage".[10] This viewpoint is strongly supported by the authors, who provide numerous examples of its ongoing development throughout the African continent.

African governments must enact comprehensive national legislation governing drone operations. This legislation should outline the legal parameters controlling the use of drones by government entities, private individuals, and non-state actors. Furthermore, it must establish regulations concerning the permissible locations and methods for drone deployment, particularly in sensitive areas, while imposing stringent registration and licensing requirements. Establishing clear standards and accountability measures will reduce the risk of drone misuse by state, private, and non-state actors. This, in turn, should boost public trust in government-approved drone operations. Furthermore, regulatory clarity will support the secure incorporation of drones into civilian industries, such as agriculture and healthcare.

At present, the degree to which the policy frameworks throughout Africa effectively facilitate strategic decision-making and threat management pertaining to drones remains uncertain. South Africa represents a unique case in this regard, having instituted a National Cybersecurity Policy Framework (NCPF) in 2015. This framework was specifically developed to address challenges such as fragmented policymaking, inadequate coordination of responses, and the lack of international collaboration and public–private partnerships.[11] However, the extent to which the NCPF has contributed to enhancing strategic decision-making and threat management concerning drones remains ambiguous.[12] Despite the establishment of the

[10] Meyer, J. and Haafizah, K., "Drone operations in South Africa: An overview of the legal landscape", *Cliffe Dekker Hofmeyr*, 19 June 2024, www.cliffedekkerhofmeyr.com/news/publications/2024/Practice/Corporate/corporate-commercial-alert-19-June-2024-drone-operations-in-south-africa-an-overview-of-the-legal-landscape.

[11] Van Nieuwkerk, A., "The Impact of Digital-Driven Warfare on Africa", *The Thinker* 90, no. 1 (2022): 20–27.

[12] Sutherland, E., "Governance of Cybersecurity—The Case of South Africa", *The African Journal of Information and Communic*ation 20 (2017): 83–112.

African Drone Forum,[13] which aims to leverage drone technology across Africa for the improvement of medical supply chains, the connection of isolated regions to commercial opportunities, and assistance in disaster management—there exists a paucity of published evidence indicating that African policymakers are actively involved in shaping effective drone policies within the African Union (AU) framework embodied by the African Peace and Security Architecture (APSA) and the African Governance Architecture (AGA).

What is evident is a bottom-up approach to regulating the space that drones are and will occupy in the future. A comprehensive database of regulations governing the use and licensing of operations and piloting can be found at the online portal Global Drone Regulation Database.[14] This online portal provides an interactive world map with active links to country-specific regulations and/or the authority webpages where those regulations can be found. Listing every regulation per country is beyond the scope of this chapter, as there are 54 countries on the African continent. There are, however, countries with little or no information about drones' regulatory frameworks and associated governing bodies, including Sudan, South Sudan, Eritrea, Somalia, Ethiopia, the Central African Republic, the Republic of Congo, Equatorial Guinea, Angola, Guinea, and Lesotho.[15] Considering these countries, there seems to be a regulatory void in the northeast of the continent—an area known as a notorious conflict zone in Africa. Other regions that are distinctly linked to significant terrorism, such as West Africa and Northern Africa, appear to have started the arduous task of regulating the use of drones and drone-related industries. An intriguing avenue for future research is to explore how each country regulates its airspace and industry, and to what extent they adhere to international regulatory frameworks, such as the International Civil Aviation Organization's (ICAO)[16] regulations for drones and those published by the European Union Aviation Safety Agency (EASA).[17]

The distinct geographical and socio-economic issues across different regions of Africa necessitate tailored regulatory measures for drone operations. It might also initiate the development of unique technologies to cope with Africa's harsh environmental realities, including urban metropoles, rugged coastal environments, deserts, and rainforests. Furthermore, the areas surrounding the African continent experience multilateral interaction and utility, such as the shipping routes around the Horn of Africa, through the Mozambique Channel, and around the southern tip of Africa. Regulating for these areas will result in complex regulatory frameworks and memoranda of understanding focussed on onshore, coastal, and offshore drone

[13] African Drone Forum, "About ADF", www.africandroneforum.org/.
[14] Global Drone Regulation Database, 2024, www.droneregulations.info/South+Africa/ZA.html#country-search.
[15] Ibid.
[16] International Civil Aviation Organization (ICAO), www.icao.int/about-icao/Pages/default.aspx.
[17] European Union Aviation Safety Agency (EASA). Drones—National Aviation Authorities resources, www.easa.europa.eu/en/light/topics/drones-national-aviation-authorities-resources.

use by diverse user groups. Additionally, maritime drones (capable of aerial, surface, and sub-surface operations as outlined in Chap. 1 of this volume) are typically used in the offshore energy sector and for shipping maintenance and hydrography. These regulations must consider the complexities of drone use in busy waterways around coastal areas and drones that are or will be multi-domain capable, meaning they can operate in both airspace and aquatic environments (surface and sub-surface). Denys Riva and Tshegofatso Ramachela, researchers for the Institute for Security Studies, speculated about these developments already in 2022.[18] As such, Meyer and Haafizah state that a comparison of global regulations reveal that "South Africa's approach shares similarities with many countries. Common themes include proximity to airports, weight limits and visual line of sight requirements. However, significant differences emerge in registration and licensing requirements, daylight flying rules and restrictions on flying over national parks. In Africa, drone regulations vary widely. Countries like Morocco and Egypt have embraced drone technology with more relaxed regulations, while others have imposed strict bans or restrictive rules. South Africa's regulations strike a balance between innovation and safety, incorporating privacy concerns into the legal framework."[19]

These complexities will require experts from the field (academia, users, industry and decision-makers) to be involved in the formulation of national and international policy frameworks to regulate this type of technology—i.e., a bottom-up approach. Bureaucrats can enhance this development process with hands-on interaction in work groups to ensure early alignment with governmental processes and procedures. A notable example of how this is accomplished is the Wassenaar Arrangement (WA), which serves as a multilateral governing body for dual-use and military technology. The WA facilitates the interaction of government officials from various states and experts in different industries to co-create international governance of these technologies and knowledge. Drone technologies, spanning from component to system levels, are regulated under the Dual-use Control List, identified by terms like "unmanned," rather than "drones," in Category 8 and within the Munitions List.[20] This is an excellent example of how international regulations are compiled in multilateral settings. South Africa is a member of the WA and therefore has expertise that could be shared with the broader African community of drone users on how to establish governing forums, potentially within the scope of the AU, to draft and maintain drone-related policy frameworks for Africa that also align with global standards.

[18] Riva D. and Ramachela T., "Can a new wave of drone tech make Africa's seas safer?", *Institute for Security Studies*, 30 November 2022, https://issafrica.org/iss-today/can-a-new-wave-of-drone-tech-make-africas-seas-safer.

[19] Meyer and Haafizah, "Drone operations in South Africa: An overview of the legal landscape", 19 June 2024.

[20] The Wassenaar Arrangement, "List of Dual-Use Goods and Technologies and Munitions List Compiled by the Wassenaar Arrangement Secretariat 2023", Wassenaar Arrangement, www.wassenaar.org/app/uploads/2023/12/List-of-Dual-Use-Goods-and-Technologies-Munitions-List-2023-1.pdf.

That said, an attempt at this type of multilateral policy formulation is visible at the AU and the NEPAD High-Level Panel on Emerging Technologies (APET) document titled—"Drones on the Horizon—transforming Africa's Agriculture".[21] Recommendations from this APET report aim to guide the adoption of drone (albeit referred to as UAV) technology across Africa, focusing on both national and continental actions. While the report focuses on agriculture, it also includes several cross-cutting recommendations for the use of drones in Africa. These are:

- At the national level, the Panel recommends that governments first evaluate the opportunity cost of UAV technology. This involves considering external factors and weighing the potential benefits, such as improved food security, enhanced healthcare, and the ability of drones to make agriculture more appealing to young people. To ensure a smooth transition, it is crucial to engage stakeholders from all sectors in discussions about UAV technology. This helps identify and address any potential resistance early on, ensuring that the introduction of drones is systematic and well-received.
- Public awareness is another key focus. There should be widespread campaigns aimed at educating the public about UAVs, emphasising the distinction between their civil and military applications. This will help alleviate concerns and improve public acceptance. Safety, security, and privacy issues should also be addressed within these campaigns to ensure that the public understands the regulations and safeguards in place.
- Governments are also encouraged to tackle the financial and technical barriers to UAV adoption. This could involve offering subsidies, supporting small and medium-sized enterprises, or fostering the development of cooperatives. Additionally, creating a robust governance framework for drones, including clear regulations on licensing and registration, will help facilitate broader adoption. Public–private partnerships should be actively promoted to drive the uptake of UAV technology, as these collaborations can play a crucial role in scaling the technology across sectors.
- National regulatory frameworks for UAVs should strike a balance between addressing public security concerns and promoting innovation, economic development, and youth entrepreneurship. National Civil Aviation Authorities should be tasked with developing regulatory environments that enable the use of UAVs, particularly in precision agriculture.
- To build a sustainable drone ecosystem, governments should invest in R&D and capacity building. This includes funding initiatives to train licensed drone pilots, scientists, and regulators, helping to create a skilled workforce capable of supporting the growing UAV sector. For smallholder farmers, the Panel recommends encouraging crop intensification by supporting the planting of crops in contiguous areas. This would help create larger, more efficient holdings that can

[21] AU and NEPAD, "Drones on the horizon—transforming Africa's agriculture", High Level APET Report, 8 June 2018, www.nepad.org/file-download/download/public/115553.

fully leverage UAV technology for precision agriculture, improving productivity and sustainability.

- At the continental level, the AU should lead the development of a unified regulatory framework for UAVs, harmonising policies across the continent. This will ensure consistency and effectiveness in the implementation of drone technologies in different countries and regions. Additionally, fostering South-South collaborations and regional partnerships will be essential for sharing knowledge, best practices, and resources to accelerate the adoption and scaling of UAV technology across Africa. These recommendations collectively aim to create an enabling environment for the widespread use of UAVs, particularly in agriculture, with the potential to drive significant improvements in food security, economic development, and innovation across the continent.[22]

These examples show that Africa is gradually developing drone regulations, moving from expert insights upwards and bureaucratic frameworks downwards. Nevertheless, significant work remains to be done. This work is critical to ensuring that current and future cooperative security and defence relations involving the use of drones are well-regulated, enhancing trust in local authorities. Ethical standards and accountability are pivotal to such trust.

1.3 Promote Ethical Standards and Accountability in Drone Warfare

> there remains a need for robust transparency and accountability mechanisms in military operations, both for African States and third party States [sic] using armed drones on African soil[23]

African governments must establish clear and transparent regulations for the use of armed drones, focusing on proportionality, precision, and civilian safety. This evolving regulatory framework was previously addressed. Creating oversight bodies, like independent drone councils, can help ensure these ethical guidelines are followed in drone operations. Upholding ethical standards in drone warfare enhances the legitimacy of African governments' use of drones, both domestically and internationally. Well-defined accountability measures can help decrease civilian casualties, alleviate public concerns, gain support for the use of military drones when necessary, and reduce political repercussions from mishaps during drone operations.

It is important to stress that African nations do not exclusively operate military drones on the continent. Western countries, including the US and France, also

[22] Ibid.

[23] PAX, "Dawn of the Drones over Africa", 3 June 2022, https://paxforpeace.nl/news/dawn-of-the-drones-over-africa/.

contribute to the deployment of armed drones in African conflict zones, incorporating their own interpretations of ethical drone usage. Moreover, non-state actors such as Boko Haram and various terrorist organisations employ drones to achieve their malicious objectives. These conflict areas present lucrative opportunities for drone supply chains, primarily coordinated by manufacturers and exporters from China, Turkey, the United Arab Emirates, and Iran.[24] This complex user environment creates a wicked dilemma for developing ethical standards, driven by the question—'who's ethics'?

1.4 Regional Cooperation, Foreign Partnerships and Dependency

African states, particularly those facing cross-border threats, should establish cooperative frameworks for drone joint operations (inclusive of surveillance and intelligence sharing—but not limited to these). This may include the development of a regional drone command center or a shared airspace monitoring system through organisations such as the AU or Regional Economic Communities (RECs). For example, in 1981, the Protocol Relating to Mutual Assistance in Defence was established between the governments of the member states of the Economic Community of West African States (ECOWAS).[25] This type of regional cooperation will allow for cooperative (or regional) drone operations in the interest of all parties involved.

The Southern African Development Community (SADC), another example of a REC, also includes a Mutual Defence Pact ratified in 2003.[26] While the SADC MDP 2003 lacks specific provisions for joint surveillance, it does encompass defence-related measures broadly. Joint drone operations between states should be feasible under the MDP 2003, provided an agreement is established regarding their precise use and designated airspace. This arrangement is usually detailed in rules of engagement, which outline limitations on methods, equipment, and duration. Nevertheless, there are general provisions for joint training, intelligence sharing, and collaborative technology development that allow for a potential review of this defence pact. This review could be updated to incorporate contemporary advancements in technology, including drones, artificial intelligence, and the integration of both technologies. Then there is also Article 10 (Supplementary Agreements) stating that "State Parties may, in respect of any particular issue covered by the provisions of this Pact, make such subsequent agreements, of a specific or general nature, as would, in their opinion, enhance the effective implementation of

---

[24] Ibid.

[25] Protocol Relating to Mutual Assistance in Defence, 1981, https://amaniafrica-et.org/wp-content/uploads/2021/04/Protocol-Relating-to-Mutual-Assistance-in-Defence-1981.pdf.

[26] SADC, "SADC Mutual Defence Pact 2003", Southern African Development Community, www.sadc.int/document/sadc-mutual-defence-pact-2003 and www.sadc.int/sites/default/files/2021-08/SADC_Mutual_Defence_Pact2003.pdf.

this Pact".[27] One future supplementary agreement could address the use of drones for cross-border surveillance and intelligence gathering aimed at mutual defence against threats typically associated with hybrid warfare.

The SADC MDP 2003 is an example of a bottom-up approach to creating cooperation on the African continent. This was followed in 2009 by the AU Non-Aggression and Common Defence Pact (2009).[28] The AUNCDP 2009 has broad provisions for cooperation and common defence across the African continent, aimed at creating stability—as typically found in Article 2. The Pact does not address specific technology. Once a common understanding is reached among bilateral and multilateral African partners, there should be no reason why drones cannot be used for surveillance to enhance regional defence, security, and stability. This is also permissible under the African Charter on Maritime Security and Safety and Development in Africa (Lomé Charter) of 2019, which allows for surveillance using maritime drones (air, surface, and subsurface).

The latest edition to mutual defence pacts on the African continent is the Sahel Security Alliance (AES) consisting of Mali, Niger and Burkina Faso[29]—aiming at defence and security cooperation—typically in response to rising tides of terrorism in the region. This would also be a perfect vehicle for cross-border drone surveillance operations. The only obstacles to cooperative surveillance, defence and security arrangements on the African continent (and anywhere else in the world, for that matter) are trust, frictional national interests, and lack of agreement between states on the modalities and scope of such operations.[30]

Enhanced regional collaboration can address cross-border security challenges more effectively and give African states greater leverage in negotiations with foreign entities. A unified position and shared infrastructure diminish redundancy, improve coordination, and fortify regional security against state and non-state threats.

For cooperation and collaboration to be effective, diverse and balanced foreign partnerships are required to avoid dependency, as the dependency theory suggests. African states should pursue partnerships with various foreign suppliers to mitigate the risk of excessive dependence on a single country or supplier and avoid 'drone colonialism.' This might be easier said than done. Countries such as South Africa, with a rich history of technology development in their defence sector, will avoid drone colonialism much easier than countries such as Eritrea, which is a net

---

[27] Ibid, 4.

[28] African Union, "African Union Non-Aggression and Common Defence Pact", 18 December 2009, https://au.int/sites/default/files/treaties/37292-treaty-0031_-_african_union_non-aggression_and_common_defence_pact_e.pdf.

[29] Al Jazeera, "Mali, Niger and Burkina Faso establish Sahel security alliance", *Al Jazeera*, 16 September 2023, www.aljazeera.com/news/2023/9/16/mali-niger-and-burkina-faso-establish-sahel-security-alliance and other reports.

[30] African Union, "African Charter on Maritime Security and Safety and Development in Africa (Lomé Charter)", *AU*, 26 September 2019, https://au.int/sites/default/files/treaties/37286-treaty-african_charter_on_maritime_security.pdf.

importer of defence equipment. A good example of a country that is both importing and developing drone technology is Nigeria.[31] Nigeria is importing drones from countries such as Israel and China but also has local developers and manufacturers assisted by Pakistan.[32] Then there is the battle space that attracts drone usage in Africa, "Ethiopia is thought to be the heaviest drone user of the African continent, deploying Chinese, Turkish, Emirati and Iranian drones for both surveillance and striking purposes in the civil war against various armed groups in Tigray and Oromo."[33] These conditions quickly create dependencies and can be exploited through proxy groups or cover action programmes to ensure instability supporting a fertile export market.

By negotiating equitable contracts that include technology transfer provisions, these states can progressively develop the necessary skills and technology for autonomous drone operations. A diversified approach ensures African governments maintain authority over their defence decisions, reducing susceptibility to geopolitical influences that might stem from regional and/or international collaboration. Furthermore, technology transfer provisions can expedite the enhancement of local capabilities, allowing countries to ultimately produce and sustain their own drone fleets whilst cooperating with international partners on favourable and sustainable terms.

1.5 Enhance Cybersecurity and Non-kinetic Counter-Drone Capabilities

African governments are urged to invest more in cybersecurity measures to safeguard their drones (i.e. the ability to maintain control inflight) and the associated data from hacking and electronic interference. Furthermore, developing anti-drone technologies, such as signal jamming or interception systems, are essential to counter the potential threats posed by non-state actors utilising drones. This discussion goes hand-in-hand with the previous point on establishing and sustaining an indigenous drone technology industry. Africa will likely leapfrog to counter-drone technology due to the limited R&D investment in military technology development on the continent. This is especially true as the rest of the developed world still grapples with developing counter-drone technology. The growth of the drone industry for military purposes, along with dual-use systems converted for military use, contributes to counter-drone technology evolving into a significant market

[31] Military Africa, "Taking Flight: Exploring the Landscape of Drone Manufacturers in Nigeria", *Military Africa*, 3 February 2024, www.military.africa/2024/02/drone-uav-manufacturers-in-nigeria/.

[32] PAX, "Dawn of the Drones over Africa", 3 June 2022, and Military Africa, "Nigerian military drones (UAV)—a complete list", 13 December 2023, www.military.africa/2020/10/nigerian-military-drones-uav-current-army-air-force-and-navy-uav-inventory/.

[33] Ibid.

shrouded in secrecy. These technologies will also be closely monitored under the dual-use control lists of the Wassenaar Arrangement export regulations.

By strengthening cybersecurity and counter-drone capabilities, African governments can effectively protect critical data and defend against hostile drones employed by state actors and/or non-state actors (insurgent and/or terrorist groups). These measures are vital to ensuring that drone technology acts as a force multiplier in the national interest rather than becoming a security liability.

1.6 Expand Applications of Drones Beyond Military Use for Civilian Benefit

Governments should increasingly support the use of drones in civilian sectors, such as agriculture, disaster response, environmental monitoring, and public health. Incentives for private sector and NGO involvement in drone technology can promote wide-ranging societal benefits. Encouraging drone use for non-military purposes demonstrates the positive potential of drones, fosters innovation, and improves public perception. For instance, using drones in health logistics (e.g., medical deliveries to remote areas) could contribute to national development goals and increase government support for drone technology. These initiatives are currently visible when considering the AU Executive Council Decision EX. CL/Dec. 986–1007 (XXXII) requesting the "AU and Member States to harness drones for agriculture as one of three emerging technologies of relevance for African development".[34] This has typically resulted in the document discussed earlier—Drones on the Horizon—transforming Africa's agriculture (AU and NEPAD High Level APET Report.[35]

Similar examples can be found for medical and humanitarian aid, where drones are operationalised for the common good of societies. One such initiative is called VillageReach, "Establishing drone networks requires time, investment, and coordination with ministries of health and civil aviation authorities. VillageReach has helped start drone programs in the Democratic Republic of Congo (DRC), Malawi, and Mozambique, and now works in more than 20 countries."[36] A key aspect of such initiatives, within the context of the current dependence of Africa on foreign drone suppliers still, is partnering with international companies. VillageReach is partnered with Australian and British drone industries to deliver their services. Other international partners in the funding and collaborating on the African continent within the context of medical and humanitarian aid programmes and typically

[34] FAOUN, "African Union embraces drones technology for agriculture", Food and Agriculture Organisation of the United Nations, 5 February 2018, www.fao.org/e-agriculture/news/african-union-embraces-drones-technology-agriculture.

[35] AU and NEPAD, "Drones on the horizon—transforming Africa's agriculture", High Level APET Report, 8 June 2018.

[36] Leedom, M., "Drones Deliver Humanitarian Aid in Africa", *Thinking Global Health*, 19 June 2024, www.thinkglobalhealth.org/article/drones-deliver-humanitarian-aid-africa.

where governments are absent from support, are non-governmental groups such as the Melinda and Bill Gates Foundation and "donor agencies like USAID, UKAID, GIZ, DANIDA, NORAD, CIDA".[37]

Then, there is the protection of wildlife heritage for certain nations. These are obviously linked to the sustainable development goals of the UN, but more closely to the socio-economic well-being of specific countries. Use cases can be found in Sub-Saharan Africa, where rhinoceros populations (both black and white species) are under constant threat from poachers.[38] Similarly, drones are used for conservation in countries such as Malawi (in partnership with the Chinese government)[39] and Uganda (protecting lions).[40] In South Africa, the Endangered Wildlife Trust has been operating drones within the context of nature conservation since 2017. Their broad uses are categorised as:

- "Surveys for birds, bird nests, and mammals for conservation purposes.
- Topographical and vegetation surveys.
- Provide aerial support to conservation teams during operations.
- Assist authorities in locating injured animals, carcasses, poisoning, and poaching incidents.
- Inspect and photograph electrical infrastructure for maintenance and survey purposes.
- Use RPAS to elevate telemetry antenna to locate wildlife fitted with tracking devices.
- Attach anti-collision devices to linear electrical infrastructure (using a system developed with the help of Eskom-Research, Development and Testing)
- Operational assistance during electrical infrastructure maintenance.
- Aerial photography and videography.
- Assisting specialists with surveys relating to Environmental Impact Assessments and related audits."[41]

As stated earlier, this dependency should be mitigated over time by establishing domestic drone development and manufacturing capabilities.

[37] Ameso, E.A. and Mwonzora, G., "Medical Drones in Africa: A Gamechanger for the Continent's 'Ailing' Health Sector?", *European Association of Development Research and Training Institutes*, 9 April 2024, www.developmentresearch.eu/?p=1800.

[38] Kobilinsky, D., "Drones and AI can improve wildlife monitoring efforts in Botswana", *Wildlife.org*, 17 January 2024, https://wildlife.org/drones-and-ai-can-improve-wildlife-monitoring-efforts-in-botswana/.

[39] Droneblogger, "Malawi to expand drone use in wildlife conservation", *Drones.R.Africa*, 2 October 2023, https://dronenews.africa/malawi-to-expand-drone-use-in-wildlife-conservation/.

[40] Gray, J., "How Wildlife Researchers Use Drones to Research and Protect African Lions", *PetaPixel*, 24 September 2024, https://petapixel.com/2024/09/24/how-wildlife-researchers-use-drones-to-research-and-protect-african-lions/.

[41] Pretorius, M., "Technology taking conservation to new heights", *Endangered Wildlife Trust*, circa 2021, https://ewt.org.za/fs-feb-2021-technology-taking-conservation-to-new-heights/.

1.7 Encourage International Norms and African-Led Dialogue on Responsible Drone Use

African states should actively contribute to shaping international dialogues concerning drone ethics, utilisation, and regulation within established forums such as the UN, the EU and/or the AU. Promoting African-led norms on the responsible use of drones may also encompass the proposal of new frameworks specifically tailored to the African context. Contributions from African countries to global drone governance will ensure that African viewpoints and security requirements are duly considered in international standards. Developing shared norms will likewise assist African countries in addressing challenges pertinent to sovereignty, ethical usage, and external influences in drone warfare.

2 Conclusion

The policy measures outlined above prioritise self-sufficiency, ethical governance, and regional resilience. By implementing these recommendations, African governments can effectively leverage drone technology to enhance security measures and civilian applications while maintaining sovereignty and stability. Fundamentally, these initiatives will enable African states to utilise drones as tools for empowerment and economic growth, thereby fostering a secure, autonomous, and ethically responsible future in the age of unmanned systems.

This book aimed to promote a deeper understanding of drones within African battlespaces by grounding the discussion in African realities—geopolitical dynamics, non-state actor involvement, ethical considerations, and logistical constraints. We believe we have made small strides to advance the field by providing nuanced, Africa-specific insights that respect and respond to Africa's unique but complex social, political and economic landscape. As drones continue to redefine modern warfare and security, the African experience offers a critical lens through which to examine the broader implications of this transformative technology and the respective operative applications in the 21st-century battlespace.

Wayne Stephen Coetzee is a senior lecturer in Political Science and former Director of the International Programme in Politics and Economics at University West in Sweden. Since 2018, Coetzee has been a visiting scholar at the University of the Western Cape, South Africa. His research primarily focuses on security politics, emphasising the interactions between public and private security, transnational defence collaborations, foreign and security policy, and the global arms trade. His publications appear in international outlets such as Cambridge University Press, *Defence & Security Analysis*, and *Cooperation and Conflict*. Over the past decade, Coetzee has also actively participated in public lectures and debates, authored "popular science" articles, and provided media insights on various subjects, including the Ukraine-Russia War, US-China relations, international law, and the security risks of environmental degradation. Coetzee earned his PhD from the School of Global Studies at the University of Gothenburg and holds MA and BA degrees from Stellenbosch University.

Dries Putter is a senior lecturer at the Faculty of Military Science (FMS) at Stellenbosch University. He has been an active-duty senior naval and intelligence officer in the South African National Defence Force since 1988. As the Chair for Security and Africa Studies at the FMS, his interests include intelligence studies, specifically counterintelligence, grey zones, and hybrid warfare. Putter is the Chair of the FMS Research Committee and a member of the Internationalisation Committee. He is part of the Steering Committee for the newly established Defence Artificial Intelligence Research Unit (DAIRU). He is also a researcher for SIGLA at the FMS (Stellenbosch University), an affiliated member of the National Security Hub (University of Canberra), and a researcher for the Africa Research Institute, Óbuda University (Hungary). His research interests encompass intelligence, counterintelligence, covert action, hybrid warfare, arms control, and the defence industry, with publications in the *International Journal of Intelligence and Counterintelligence, Global Policy, and Scientia Militaria.*

The manufacturer's authorised representative in the EU is Springer Nature Customer Service Centre GmbH, Europaplatz 3, 69115 Heidelberg, Germany. If you have any concerns regarding our products, please contact ProductSafety@springernature.com

Printed and bound by CPI Group (UK) Ltd, Croydon, CR0 4YY
30/09/2026
02225574-0015